# THE POLITICS OF BIOTECHNOLOGY IN NORTH AMERICA AND EUROPE

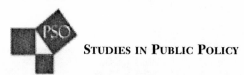

**STUDIES IN PUBLIC POLICY**

**Series Editor: Paul J. Rich, Policy Studies Organization**

Lexington Books and the Policy Studies Organization's **Studies in Public Policy** series brings together the very best in new and original scholarship, spanning the range of global policy questions. Its multi-disciplinary texts combine penetrating analysis of policy formulation at the macro level with innovative and practical solutions for policy implementation. The books provide the political and social scientist with the latest academic research and the policy maker with effective tools to tackle the most pressing issues faced by government today. Not least, the books are invaluable resources for teaching public policy. For ideas about curriculum use, visit www.ip-sonet.org.

*Analyzing National and International Policy: Theory, Method, and Case Studies*, by Laure Paquette

*Developmental Policy and the State: The European Union, East Asia, and the Caribbean*, by Nikolaos Karagiannis

*Policymaking and Democracy: A Multinational Anthology*

*Policymaking and Prosperity: A Multinational Anthology*

*Policymaking and Peace: A Multinational Anthology*
    Three-volume set edited by Stuart Nagel

*Equity in the Workplace: Gendering Workplace Policy Analysis*, edited by Heidi Gottfried and Laura Reese

*Politics, Institutions, and Fiscal Policy: Deficits and Surpluses in Federated States*, edited by Louis M. Imbeau and François Pétry

*Innovation and Entrepreneurship in State and Local Government*, edited by Michael Harris and Rhonda Kinney

*Comparative Bureaucratic Systems*, edited by Krishna K. Tummala

*Public Policy in Israel: Perspectives and Practices*, edited by Dani Korn

*The Struggle of Soviet Jewry in American Politics: Israel versus the American Jewish Establishment*, by Fred A. Lazin

*Foreign Policy toward Cuba: Isolation or Engagement?*, edited by Michele Zebich-Knos and Heather N. Nicol

*The Work of Policy: An International Survey*, edited by H. K. Colebatch

*The Politics of Biotechnology in North America and Europe: Policy Networks, Institutions, and Internationalization*, edited by Éric Montpetit, Christine Rothmayr, and Frédéric Varone

# THE POLITICS OF BIOTECHNOLOGY IN NORTH AMERICA AND EUROPE

## Policy Networks, Institutions, and Internationalization

### Edited by Éric Montpetit, Christine Rothmayr, and Frédéric Varone

LEXINGTON BOOKS

A division of
ROWMAN & LITTLEFIELD PUBLISHERS, INC.
Lanham • Boulder • New York • Toronto • Plymouth, UK

LEXINGTON BOOKS

A division of Rowman & Littlefield Publishers, Inc.
A wholly owned subsidiary of The Rowman & Littlefield Publishing Group, Inc.
4501 Forbes Boulevard, Suite 200
Lanham, MD 20706

Estover Road
Plymouth PL6 7PY
United Kingdom

British Library Cataloguing in Publication Information Available

**Library of Congress Cataloging-in-Publication Data**
The politics of biotechnology in North America and Europe : policy networks,
institutions, and internationalization / edited by Éric Montpetit, Christine Rothmayr,
and Frédéric Varone.
    p. cm.
  Includes bibliographical references.
  ISBN-13: 978-0-7391-1247-2 (cloth : alk. paper)
  ISBN-10: 0-7391-1247-3 (cloth : alk. paper)
  ISBN-13: 978-0-7391-1248-9 (pbk. : alk. paper)
  ISBN-10: 0-7391-1248-1 (pbk. : alk. paper)
  1. Biotechnology—Government policy—United States. 2. Biotechnology—
Government policy—Canada. 3. Biotechnology—Government policy—Europe.  I.
Montpetit, Eric, 1970–  II. Rothmayr, Christine, 1968–  III. Varone, Frédéric.
  TP248.185.P65 2007
  303.48'3—dc22                                                    2006024502

Printed in the United States of America

∞™ The paper used in this publication meets the minimum requirements of
American National Standard for Information Sciences—Permanence of Paper
for Printed Library Materials, ANSI/NISO Z39.48-1992.

# CONTENTS

# ACKNOWLEDGMENTS

The idea behind this book was first laid out in Geneva, in the fall of 2002, at a meeting sponsored by the European Consortium for Political Research (ECPR). Several of the participants to this meeting had been part of the Comparative Policy Design Project, which was also conceived during an activity of the ECPR: the 1997 workshops in Berne. The Comparative Policy Design Project's main goal was to develop a better understanding of policy differences across industrialized countries. After the publication of several articles and a book on this topic, some participants in the project began wondering about differences across policy sectors. The goal of the meeting in Geneva was to identify sectors for the comparison and to clarify the end results the Comparative Policy Design Project could strive toward. We rapidly agreed that country comparisons could overestimate the importance of political institutions and underestimate that of policy networks and globalization in policy explanations. The participants agreed also that it would be better to concentrate research efforts on sectors that are not too foreign to each other. In the context of the Comparative Policy Design Project, we had already developed considerable knowledge on assisted reproductive technology—as such, we believed it wise to take advantage of that knowledge. Genetic engineering in the agri-food sector (known as GMOs) imposed itself as an appropriate second sector. On the one hand, just like assisted reproductive technology, GMOs belong to the wider areas of biotechnology and the knowledge economy. On the other hand, unlike assisted reproductive technology, GMOs are produced by large firms for international markets. These differences appeared important to us given our goal to shed light on the policy effect of networks and globalization.

Since Geneva, we have spent considerable time gathering information on the evolution of these two sectors in nine countries. We've also had several occasions to share ideas and agree on a common analytical framework. This book is the main result of these four years of research. During these four years, the contributors to this book received financial support from several organizations. We acknowledge the financial support of the Social Sciences and Humanities Research Council of Canada, the Fonds québécois de recherche sur la société et la culture, the Swiss National Science Foundation, the Ministère de la politique scientifique fédérale de Belgique and the Centre de recherche sur les politiques et le développement social at the Université de Montréal. We owe a special thank to the Brocher Foundation (www.borcher.ch), Geneva Switzerland, for its generous support in the final stages of the preparation of this book. Katherine Macdonald and the staff at Lexington Books were very efficient and also deserve our thanks. Audrey L'Espérance helped considerably with the index and Martin Vézina formatted the manuscript. Lastly, James Allison spent several hours, sometimes on weekends, working to improve the quality of our book. We are extremely grateful to him.

One of the original members of the Comparative Policy Design Project, Ulrich Klöti, Professor at the University of Zurich, Switzerland, passed away this year. Ulrich was an enthusiastic supporter of this project. He made a significant intellectual contribution to it and encouraged several of us to pursue our international collaboration within the framework that the project provided. We dedicate this book to his memory.

## ❶

# COMPARING BIOTECHNOLOGY POLICY IN EUROPE AND NORTH AMERICA: A THEORETICAL FRAMEWORK

*Frédéric Varone (Université de Genève),*
*Christine Rothmayr (Université de Montréal),*
*Éric Montpetit (Université de Montréal)*

The life sciences have in the past thirty years produced a steady stream of new developments, most notably in the field of biotechnology. With this advance in scientific knowledge and the development of new techniques comes the promise of great benefits: the cure of degenerative diseases, new reproductive techniques, and tools to combat malnutrition and hunger. These developments have also given rise to grave concerns over their potential negative effects, such as the return of eugenics in the form of embryo selection techniques, the contamination of the environment through genetic manipulation and a general orientation towards the commodification of life.

In the face of these controversial issues, biotechnology has been, since the late 1970s, widely discussed in the media and has become a salient topic on the political agenda (Durant et al. 1998; Gaskell and Bauer 2001; Bauer and Gaskell 2002). A number of advanced industrialized democracies have passed national legislation in response to these debates, and several supra- and international institutions have adopted norms in these fields, in particular the European Union, the World Health Organization and the World Trade Organization.

The range of public policies that regulate contentious biological issues, such as genetic and reproductive engineering, stem cell and embryo research, organ transplantation or genetically modified organisms, can collectively be labeled as *biopolicies* (Blank and Hines 2001). The field of biopolicies has been subdivided into a number of subfields; we frequently find a distinction between *green* and *red* forms of biotechnology, which refer, respectively, to biotechnology in the agriculture and food sector versus biotechnology in human and veterinary medicine, including pharmaceutical sciences. We prefer to distinguish between *human* and *nonhuman* applications of biotechnology. This edited volume compares the design of public policies for both fields of biotechnology by analyzing two specific sectors in nine countries on two continents: on the one hand, policies addressing *genetically modified organisms in the agro-food sector*, and on the other hand, policies for *assisted reproductive technologies*.

The study includes six EU member states (France, Great Britain, Germany, the Netherlands, Sweden and Belgium), two North American countries (USA and Canada), and one European but non-EU-member state (Switzerland). Through the use of these country case studies and a subsequent comparative analysis, we identify the differences and commonalities for these eighteen sector- and country-specific biotechnology policies. We then provide an explanation for the differences and similarities across the nine countries and two policy sectors.

*Assisted reproductive technologies* (hereafter ART), our first policy field, fall into the category of red, or human, biotechnology. We understand ART to consist of those techniques where egg and sperm are not united through, or an embryo is not created through, sexual intercourse but rather through medical intervention. The invention of the technique of in vitro fertilization (IVF) in the late seventies and its routinization in the mid-eighties created the basis for new medical techniques in assisted reproductive technologies, such as intracytoplasmic sperm injection (ICSI), preimplantation diagnostics (PID), genetic screening and engineering, as well as new research fields, such as research on embryos, on stem cells, and both therapeutic and reproductive cloning. Our second policy field, *genetically modified organisms* (GMO), is situated in the realm of green, or nonhuman, biotechnologies. A GMO is an organism whose genetic composition has been altered through a laboratory gene transfer. In this research project, we are exclusively interested in GMOs in the *agro-food sector*,[1] which are those GMOs that have been designed for the purposes of entering into the (human) food chain through agriculture and genetically modified food (GM food).

Both fields raise a number of questions in terms of expected benefits and potential risks, which has produced varying degrees of controversy in Europe and North America. Opinion polls have revealed that the social acceptability of green and red biotechnology varies considerably across countries and also across types of applications. The Eurobarometer study 52.1

"The Europeans and biotechnology,"[2] conducted in 1999, revealed a high acceptance of red, health-related biotechnology, while showing a far greater and growing skepticism towards green, agricultural biotechnology. This skepticism towards biotechnology focused, however, on particular applications and not on the technology as such. While the vast majority rejected genetically modified food (novel food), most people assessed the detection of hereditary diseases as a good and morally acceptable purpose of biotechnology. Other medical applications, however, were regarded as morally problematic, such as cloning and research on human embryos, or the use of genetic diagnosis for preimplantation diagnosis. Furthermore, public opinion in the United States and Europe differs and we find a greater resistance against green biotechnology in Europe than in North America (Gaskell, Thompson, and Allum 2002).

In addition to the variation of public opinion across countries and types of application, we also observe differences in the market shares and economic importance of biotechnologies. First of all, there is considerable variation in the availability of ART treatments across countries. The European IVF-monitoring program (EIM) of the European Society of Human Reproduction and Embryology (ESHRE) is the only source for comparative data on ART treatment (Nygren and Andersen 2002).[3] The data shows that the percentage of ART infants as a percentage of all national births varies considerably among countries from about 1.4 percent in France to 3.6 percent in Iceland. The number of IVF cycles per one thousand women aged fifteen through forty-nine also varies considerably from about 2.6 in Switzerland to 7.3 in Denmark. There are additional discrepancies among Western countries in terms of embryo research and research on stem cell lines: in the USA, there are a large number of embryonic stem cell lines in use (about sixty); in Great Britain, the first stem cell line was derived only recently; in Germany, nine projects using imported stem cells were authorized as of March 2006, since derivation within Germany is prohibited. Recent reports also indicate that the number of businesses in the field of stem cell research has decreased in the United States because of the uncertain legal context, but in other countries, in particular Sweden, the number of businesses in this research sector is increasing dramatically (Red Herring 2003).

While embryonic stem cell research is a growing, but not-yet-profitable market, the development of GM crops has become a lucrative business. Since the mid 1990s the global GM crop area has grown steadily. By 2002, 5.5 to 6 million farmers in sixteen countries had planted 58.7 million hectares of GM cultivars, producing an area equivalent to half the land area of the United Kingdom. Four countries grow 99 percent of the area of global transgenic crops: the United States, Argentina, Canada and China. Soy, maize and colza are the three main GM crops. The two most common traits induced in cultivars are tolerance to herbicides (about 75 percent of all GM cultivars) and tolerance to pests (ISAAA 2003). European countries are not particularly amenable to the use of GM crops, despite the European Commission's strategy plan that seeks to promote the life sciences and biotechnology, and to encourage the European Union to become a leading knowledge economy by 2010 (European Commission 2002). If we examine investments in research and development in GMO technologies, however, by considering, for example, the relative impact of publications in the biotechnology field or public funding of R&D activities in biotechnology, several of the countries that are insignificant players in the growing of GM crops and in GM food markets appear to play an important role in the development of GMOs. In 2000, Belgium spent the highest percentage of its R&D budget on biotechnology (13.8 percent) and was, by the end of the 1990s, the United States' main partner for biotechnology, accounting for almost a quarter of all American biotechnology imports and exports (Van Beuzekom 2001: 22, 36). Furthermore, Switzerland, Finland, the Netherlands, Sweden and Germany had, like the United States, above average impact rates for biotechnology publications from 1986–1998. Small countries like Switzerland and Ireland are also important biotechnology import partners for the United States; Switzerland had an 11.7 percent share of American biotechnology imports from OECD countries in 1999, and Ireland had 6.9 percent (Van Beuzekom 2001: 16, 36). Thus, the commercial interests of countries vary according to the subfield of biotechnology that we are studying, and countries which are not necessarily big players in the GMO agro-food business may have considerable economic interests in other subfields, namely in research and development and in red rather than green biotechnology.

This short overview indicates that it is of particular interest to compare both dimensions of biopolicies, the human and the nonhuman dimension. Policies for the use of GMO in the agro-food sector vary across North America, the European Union and other countries; the growing economic importance of GMOs in the agro-food sector has led to a trade conflict between the United States and Canada, on one side, and the EU on the other (Bernauer and Meins 2003). There is, however, also considerable variation among EU countries in terms of the concrete implementation of EU directives on the contained use and deliberate release of GMOs. The pattern of policies for red biotechnologies, and in particular for ART, is even less clear (see Bleiklie, Goggin, Rothmayr 2004; Varone, Rothmayr, Montpetit 2005) as there is considerable variation within the EU, as well as between Canada and the United States. The United Kingdom, for example, has encouraged embryo research with fairly liberal policies, but Germany adopted very restrictive policies in order to prohibit all forms of invasive embryo research. Preimplantation diagnostics are accepted in France, but they are prohibited in Switzerland. While the American Congress has thus far failed to adopt a comprehensive act to promote an ethical use of embryos in research, the Canadian House of Commons adopted such a law in 2004. *Why then are some countries comparably more restrictive when it comes to embryonic stem cell research, for example, but much more liberal when it comes to GM food? Why have some countries adopted policies promoting biotechnology research and its applications, while others impose considerable regulatory restrictions on both ART and GMO?*

There are currently no comparative studies analyzing public policies for green and red biotechnologies in Europe and North America, which we would need in order to answer these types of questions. This volume intends to contribute to closing this gap by describing, comparing and explaining policy choices in green and red biotechnology. Our contribution as political scientists to the ongoing debates on the regulations of biotechnology consists of examining *"how" and "why" governments pursue particular courses of action or inaction in the fields of ART and the GMO agro-food sector* (Heidenheimer et al. 1990). Hence, the contribution of this volume is twofold. First of all, we aim to provide a broader empirical base of knowledge on how public policies are designed and chosen in the two fields. Despite the considerable amount of political activity and the public attention surrounding these biotechnology issues, there is a lack of political science research on this topic (Rothmayr and Varone 2002). This seems particularly evident if we compare what has already been undertaken in the growing fields of bioethics, biolaw, and bioeconomics to the still missing—or at least underdeveloped—political science research in biopolitics.

We intend, secondly, to contribute to theory development in the field of comparative public policy. Policy analysts are increasingly confronted with new issues that no longer fit into clearly drawn policy sectors within isolated territorial jurisdictions. The issues that are now emerging from the knowledge economy of scientifically advanced countries tend to overlap several traditional sectors. The opposing processes of decentralization and globalization are redefining territorial jurisdictions. This phenomenon is creating new opportunities, but also new constraints; we are observing new forms of collaboration among actors, but also new tensions. In contrast to existing policy books, our objective is to account for this new and transformed policy-making context by examining a wide range of interrelated variables drawn for older and more recent approaches. Specifically, we assess the policy effect of variables derived from three competing approaches to policy analysis. We have labeled these approaches as follows: the policy network, the country-pattern and the internationalization approaches. The originality of our perspective rests with its integration of a bottom-up perspective that suggests that issues emerge, and policy choices are made, in the context of specific policy networks, and top-down perspectives that insist on the impact of domestic institutions and internationalization.

In the next section, we present the analytical framework in greater detail. We discuss, first, how we compare policy content across the nine countries and two sectors, and then describe the three categories of explanatory factors that have guided our empirical analysis.

## CLASSIFYING BIOPOLICIES: THREE IDEAL TYPES

Each country is (partially) unique with respect to its biotechnology policies covering ART and GMO. In other words, each country has formulated a collective problem to be solved by state intervention, has adopted its own goals, has selected its own policy instruments to achieve them, and has portrayed those who are affected by the technology in its own way. The policy design approach (e.g., Linder and Peters 1984; Bobrow and Dryzek 1987; Schneider and Ingram 1997) provides analytical concepts that enable the analysis of policies in a sufficiently detailed way and allow for a comparison across time, countries and policy fields. A policy design can, thus, be understood as a set of four attributes that permit us to distinguish one policy from another; the four attributes are: *policy goals*, understood as the intended consequences of the policy; *policy instruments*, as the means proposed to achieve the desired ends; *target groups*, as the persons whose behavior a policy intends to influence; and *implementers*, the public and/or private actors in charge of taking measures to implement the policy instruments. Finally, we also include another aspect of policy design, "policy rationales," which are the expressed justifications for the choice of goals, instruments, target groups and implementers underlying the political regulation of biotechnologies. The policy rationales link together all of the constitutive elements of the policy design (Schneider and Ingram 1997); they can be understood as the "intervention hypothesis" or "logic of action" on which state intervention is based.

In order to analyze the design of biotechnology policies, which constitutes our dependent variable, the country case studies first identify—along the traditional production chain of 'research and development,' 'experimentation,' 'production,' and 'distribution, sale and consumption'—the actors, whose behavior should be regulated by the state (i.e., the target groups) as well as the policy instruments that were implemented by state agencies in order to influence the decisions and actions of these targets groups. Scientists conducting research in private and public institutions might, for example, constitute the main target group for policies that regulate research and development in both fields, whereas the instruments used to regulate production might target farmers and physicians. Types of instruments might include bans, labeling or licensing requirements and quality controls, as well as public subsidies or reimbursement by public or private health insurances according to the policy field and phase in the production chain. This approach to policy design, as proposed by several authors (e.g., Linder and Peters 1989), is thus clearly focused on the policy instruments and the targeted groups.

After having identified and described the policy design according to the traditional policy dimensions, each case study also qualifies the overall policy design. To do so, we categorize the real world complex situations within three ideal-types of policy designs: a permissive biopolicy, an intermediate biopolicy and a restrictive biopolicy. These ideal-types are constructed to represent three univocal historical paths of technology development and regulation by the state. By definition, they are not found "*tel quel*" in reality, but they constitute a theoretical abstraction for coding and distinguishing the biopolicies implemented in various sectors and countries. According to Max Weber (1965), ideal-types of a social occurrence (as for example the content of a public policy) are useful in distinguishing between occurrences, which otherwise might be difficult to recognize in a complex reality. Only after the distinctions between biopolicies are made clear can we begin formulating hypotheses about the factors leading one country, in one specific biotechnological sector, to intervene with a pro-technology rather than choosing a position that is more cautious about technology. (See table 1.1.)

The three ideal-types of design for ART policies are distinct with respect to how strongly they protect the embryo in vitro and to what extent they adhere to a traditional notion of the family defined by genetic kinship and civil status. With respect to ART, a country with a permissive ideal-type design would typically not limit research possibilities and allow for the creation of embryos for research purposes through IVF or therapeutic cloning. Furthermore, the selection of embryos through preimplantation diagnostics

**Table 1.1. Ideal Types for the Design of Biopolicies in the ART and GMO Sectors**

| Ideal types of policy design | Empirical elements related to the ideal type for the ART sector | Empirical elements related to the ideal-type for GMO in the agro-food sector |
|---|---|---|
| Permissive ("everything is allowed with some exceptions") | - Creation of embryos for research purposes and therapeutic cloning are allowed<br>- Embryo and gamete donation are not restricted<br>- Gender selection and preimplantation diagnosis are generally permitted<br>- No limitation of access to ART (IVF, AID) in terms of civil status and sexual orientation | - Experiments in fields (deliberate release for research purposes) are generally allowed<br>- Production of GM crops (deliberate release for agricultural purpose) is generally allowed<br>- Commercialization of novel-food is generally allowed<br>- Traceability of GMO and labeling of products containing GMO are not mandatory |
| Intermediate ("several things are allowed under strict conditions") | - Research on and derivation of stem cells is allowed only on leftover embryos and therapeutic cloning is prohibited<br>- Egg, sperm, and embryo donation are permitted under specific conditions (for example, egg donation only from women undergoing IVF themselves)<br>- Genetic screening for preimplantation diagnosis is limited to predefined hereditary diseases<br>- Access to ART (IVF, AID) is limited to stable couples | - Field experiments are allowed only on a case-by-case basis (after bio-safety assessment)<br>- Production and importation of GM crops is allowed only on case-by-case basis<br>- Commercialization of novel-food is allowed only on a "case-by-case" basis<br>- Traceability and labeling of (imported) crops and products with a specific threshold of GMO are mandatory |
| Restrictive ("everything is forbidden, with some exceptions") | - Research on embryos, the derivation of stem cells and therapeutic cloning are prohibited<br>- Egg, sperm, and embryo donation are prohibited<br>- Genetic screening for preimplantation diagnosis is prohibited<br>- Access to ART (IVF, AID) is limited to heterosexual married couples | - Field experiments are de facto forbidden<br>- Production or importation of GM crops is de facto forbidden<br>- Commercialization of novel-food is de facto forbidden<br>- Traceability and labeling of imported crops and products with a specific threshold of GMO are mandatory and strongly controlled by the state |

would be permitted, embryo and gamete donation would not be restricted, and access to ART treatment would not be restricted by civil status and sexual orientation. The reverse ideal type of policy design, the restrictive design, would, to the contrary, severely limit embryo research by prohibiting the derivation of stem cells, therapeutic cloning, and all type of invasive research. The restrictive design would also prohibit the selection of embryos in vitro, prohibit embryo and gamete donation, and limit the access to ART treatment to married couples with some exceptions on a case-by-case basis. The intermediate design would limit research to leftover embryos, in other words prohibiting the creation of embryos through IVF or therapeutic cloning for research purposes. The intermediate design would also tie genetic screening and donation to specific requirements and only allow stable couples to have access to ART treatment.

The three ideal types of designs for GMOs in the agro-food sector vary with respect to how strongly the state intervenes into the areas of research, production, commercialization and distribution and, hence, how strongly the state adheres to the precautionary principle and emphasizes the protection of conventional or organic farming. In the permissive ideal-type, the deliberate release of GMOs for research and production, as well as the commercialization of novel-food, would generally be allowed; the traceability and labeling of products would not be mandatory. The intermediate type of design would require authorization on a case-by-case basis for releases, research and production, and for the commercialization of novel-food; traceability and labeling would be mandatory in order to guarantee a choice between products from biotech-agriculture and conventional or organic farming. In the case of the restrictive type, the release, production, commercialization and distribution of GMOs would not be taking place, de facto, because the respective authorizations are regularly refused.

The case studies explain, first, why the policy design in one sector corresponds to the ideal type of a permissive, intermediate, or restrictive policy design (e.g., restrictive ART regulation in Germany, permissive GMO regulation in Canada), and then discuss why the policy designs in the ART and GMO fields correspond to the same or a different ideal type of design. The concluding, comparative chapter then addresses the differences and similarities across the nine countries and two sectors based on three rough categories of variables: "policy networks," "country pattern," and "internationalization." These categories of variables are strongly related to the three traditional comparative approaches to the study of politics and policies, which Levi-Faur (2004) refers to as the *policy sector* approach (Freeman 1986; Marsh and Rhodes 1992, and all "policy networks" scholars), the *national patterns* approach (Richardson 1982, Vogel 1986, and all "comparative politics" scholars) and the *international regime* approach (Krasner 1983). Again, we have adopted the following labels to refer to these three comparative approaches: the policy network, the country pattern and the internationalization approaches.

## POLICY NETWORK APPROACH

Within the field of comparative politics, there has long been the recognition of avoiding generalizations across countries, but rather the emphasis has been on the importance of country-specific institutions and cultures. Increasingly, however, policy analysts observe policy variations, not only across countries, but also across sectors in any given country. Sectoral variations have often been imputed to policy networks.

A policy network is a group of actors, notably comprising interest groups and state agencies, who negotiate, discuss or share resources in view of contributing to the formation of the policy agenda, policy design and implementation. In this book, we use the notion of policy network rather broadly. Therefore, a single sector may consist of one or many networks (Sabatier and Jenkins-Smith 1999; Marsh and Smith 2000). The network(s) can be variously in tune with the dominant policy beliefs and ideas present in a sector (Coleman 1998). They also vary according to the degree of openness to new actors and the interconnection between civil society and the state (Bressers and O'Toole 1998). British scholars (Rhodes 1997; Marsh and Rhodes 1992) designate any network consisting of a small number of tightly interconnected actors who hold cohesive policy beliefs as *policy communities*. In contrast, they label those large networks that include a diversity of beliefs, but which are not sufficient, in themselves, to provide actors from civil society a participatory role, alongside state actors, in policy-making, as *issue networks*.

Sectors that are under the influence of policy communities should have different policies than sectors where issue networks are dominant. For instance, constructivists argue that policy problems are not given as is, but rather, they are socially constructed or framed by influential actors (e.g., Dery 1984:xi; Cobb and Elder 1983:172; Rochefort and Cobb 1994; Schön and Rein 1994). Networks certainly provide clues as to the identity of these influential actors. In a sector dominated by a policy community, it can be expected that the members of the community, holding cohesive beliefs, will have considerable influence on the definition of the nature of the problem, its causes, and in proposing workable solutions. In contrast, state actors may have more leeway to frame the problems and propose solutions in sectors dominated by issue networks. In what follows, we suggest one hypothesis regarding the influence of policy communities on biotechnology policies and one hypothesis regarding the influence of issue networks.

## Policy Communities Produce Permissive Biopolicies (H1)

The actors with the longest history in biopolicy sectors are the biotechnology industry, the research community and the medical profession. While in most countries the industry organized itself in view of influencing policy when biotechnology became economically promising in the 1970s, the research community and especially the medical profession already had powerful organizations in place to exert policy influence. It is only later, in the 1980s and the 1990s, that other groups, such as environmentalists, consumers, patients and women, began to become concerned and then emerged as new policy actors in the various sectors of biopolicy. These actors often brought new and even sometimes dissonant ideas to the sectors where industry, the research community and the medical profession had already begun policy-making work alongside state actors. In some cases, these already present actors formed policy communities powerful enough to resist any policy change demanded by emerging groups.

Why should we expect a policy community to have the capacity to resist the policy changes lobbied for by new actors in the sectoral environment? Members of industry, researchers and the medical community possess a privileged access to biotechnological knowledge and also have vested interests in this area. It is therefore unsurprising that state actors turn to this group of actors first for participating in the development of biotechnology policies or that members of this group are likely to be the first to turn to state actors to demand such policies. For this reason, and in contrast to more recent actors, the industry, researchers and the medical community had more time to develop close interconnections with the state actors who possess the prime responsibility for biotechnology policy development. These actors are also likely to develop self-regulation as a strategy of influencing and possibly preventing future state intervention. The strategy of self-regulation involves the definition of quality standards and codes of ethics within a specific sector.

In some cases, the trust that developed between the early movers and state actors became significant enough for the state actors to endorse self-regulation. As Streeck and Schmitter (1985) argue, in sectors where specific expertise is required, state actors can be convinced to accept that the groups that are the key targets of the policy play a significant regulatory role, forming what they called "private interest governments." When this occurs, the target groups have key roles in the framing of policy problems and solutions. As emphasized by Salter (2001), the policy influence of doctors in the field of biomedicine rests on their control of specialized knowledge and on the technical language they construct to communicate this knowledge (see also Hassenteufel 1997). Therefore, when a policy rests on a language that appears complex to anyone but the target groups and decisive state actors, challenges coming from outside the policy community or the self-regulatory arrangement will often suffer from insufficient credibility to encourage policy change.

In other words, a policy community, especially if it self-regulates, provides the best conditions for creating a lock-in effect. As Pierson (2000) argues, any policy choice creates increasing returns, notably through learning and coordination. A policy community rests on knowledge and a language that had to be learned by the participants, including state actors, and provide a powerful coordinative structure. Coordinative structures, Pierson (2000) explains, are valuable assets for the actors involved in them because policy-making requires collective action. Therefore, the actors of a policy community should offer strong resistance to any policy change that threatens the very existence of the policy community. In sum, if target groups form a policy community in any given biotechnology sector, we should not see, over long periods of time, any significant departure from the trajectory upon which they have initially decided.

Can we expect restrictive or permissive biopolicy decisions from such a policy community? In this book, we treat the preferences of network actors for permissive or restrictive biotechnology policy as an empirical question. This is not to say, however, that formulating hypotheses prior to conducting empirical research is impossible. As Scharpf argues, policy analysts who treat preferences as an empirical question are not

> left with the unmanageable task of having to collect empirical data on the feasible options and the specific content of idiosyncratic perceptions and preferences in every case and for every actor. Instead, researchers can make use of the same institutional information that enables the actors themselves to interact with reasonable assurance that, by and large, they will know what is going on and what to expect of each other. (Scharpf 1997: 41)

In the case of biotechnology policy, we know that some groups have a prime interest in policy-making because they are the most likely targets of policy. Whether the objective of the policy is to encourage or restrict embryonic research, the research community is the likely target: grants to encourage research will primarily benefit researchers, and regulations to restrict research will primarily target the practices of researchers. A similar logic applies to industry with regards to GMOs and to the medical community with respect to fertility treatments. When such target groups contribute to policy-making, and even more so when they are part of a self-regulatory arrangement, they will prefer the least constraining of possible policy instruments. Some level of constraint may be necessary to the extent that policy community actors are concerned about the credibility of their decisions (Carpenter 2004). As Schneider and Ingram (1997: 118) argue, however, the burden inflicted by a policy tends to be illusionary rather than real when the target of the policy is a group that either enjoys a good reputation or has resources to exert power. Potential targets in the area of biotechnology policy, including industry and scientists, are not always held in high esteem, but they possess the necessary resources to exert political influence. It is certainly uncontroversial to hypothesize that, if left to industry, researchers and the medical community, biotechnology policies are more likely to be permissive than if more recent actors, such as environmentalists and patients, participate in their formulation.

Our first hypothesis, therefore, reads as follows: *if a sector is organized around a policy community dominated by the target groups, the policy will grant a high degree of autonomy to the target groups, perhaps even supporting a self-regulatory arrangement.* In short, policy communities, understood as closed networks of civil society and state actors who share similar beliefs, should encourage the adoption of permissive biotechnology policies. A corollary of this hypothesis is that when the initial policy choices are made by a policy community, the permissive trajectory should be "sticky." Major departures from this trajectory should be difficult to accomplish, especially if the initial choice supports self-regulation.

## Issue Networks Produce Restrictive Biopolicies (H2)

As alluded to above, it is reasonable to expect policy differences across sectors, because policy networks are likely to vary from sector to sector. For a number of historical reasons, we may reasonably assume that potential target groups will, in some sectors, fail to form a cohesive policy community and therefore not be able to exert much control over the policies within this sector. We may thus ask what are the effects on policy design when a sector is characterized by the presence of issue networks. An issue network, once again, is comprised of a wide array of actors with divergent ideas and beliefs. Typically, an issue network provides a poor interconnection between civil society actors and state actors. In the area of biotechnology policy, issue networks are comprised of actors from outside the key target groups of state intervention. These actors, whom we referred to above as the *new biopolicy actors*, are often opponents to biotechnology or actors concerned about biotechnology. They include, among others, environmental groups, consumers, and women's groups. Without explicitly referring to policy networks, some authors have suggested that such groups can exert significant policy influence (see Cobb and Elder 1983; Cobb et al. 1976; Baumgartner and Jones 1993). We argue in this section that this is most likely in sectors where issue networks dominate.

It may appear counterintuitive that groups such as environmentalists are able to exert influence in the GMO policy sector. Environmental groups, like women's groups, have fewer financial resources and are much less organized than the biotechnology industry or the medical profession. In fact, the policy influence of these groups may precisely be a result of their fragmented structure. Environmental groups and women's groups are often considered as social movement organizations. In contrast to a typical interest group whose activities consist of working with or lobbying the state, a social movement organization devotes significant efforts to mobilizing wide segments of society behind a given cause. They tend to present themselves as defenders of long-term social interests. In the area of biotechnology policy, these interests include the protection of human life after conception, the rights of women over their bodies, the preservation of the environment against genetic contamination and the fight against poverty in the context of economic globalization. These causes have a large appeal and, with the help of the media, can in fact turn public opinion against target groups (Cobb and Elder 1972). In fact, social movement organizations with causes that are sufficiently appealing to mobilize a significant number of citizens can disrupt a policy community's control over a given policy (Dryzek 2000). More often than not, however, social movement organizations exert power when the network status within a sector is instead an issue network. In such a situation, the process whereby social movement organizations exert an influence is slow moving (Pierson 2004).

This type of process primarily involves the slow construction of a cognitive frame of reference, in which social movement organizations are unsatisfied with existing frames, and the emerging frame of reference provides citizens with an appealing understanding of a social problem and also offers a potential response by the state. We should recall, however, that the actors of a policy community share common interests and can thus rapidly construct a frame of reference when it does not already exist prior to undertaking any policy-making work. In contrast, social movement organizations construct cognitive frames in view of providing the entire society with an alternative understanding of a problem, which takes more time to achieve.

It should be noted that issue networks typically leave state actors with more policy-making autonomy than policy communities. As we explain in the next section, state agencies have their own policy preferences, to such an extent that we can expect policy differences to emerge when the implementation of a GMO policy, for example, is delegated to the agricultural rather than the environmental department of a country. The capacity of a department to translate its particular preference into the policy should be greater in sectors where issue networks prevail, due to the weak interconnections of these networks. State actors are also free to be more responsive to the pressure coming from policy contexts where issue networks are dominant. Rather than responding to the concerns of target groups, state actors can thus decide to adopt the restrictive policies proposed by a social movement organization that has been successful at mobilizing various actors and citizens behind a cognitive frame opposing or cautious about biotechnology.

Our second hypothesis can thus be formulated as such: *where issue networks characterize a sector, groups opposed to or concerned by potential negative effects of biotechnology can be successful at establishing an alternative understanding of biotechnology to that of target groups, and thus policies will turn out to be more restrictive.* Of course, issue networks alone are insufficient to explain the adoption of restrictive biotechnology policies, but openness of issue networks greatly encourages social mobilization activities and frees key state actors from the influence of a limited number of powerful target groups. Issue networks also encourage the adoption of restrictive policies, even in the absence of a successful mobilization by social movement organizations, when the state agency that exercises prime responsibilities in the sector has restrictive policy preferences; we discuss this element in more detail in the next section.

## COUNTRY PATTERN APPROACH

In the second set of working hypotheses, we focus on specific features of the political systems of the countries in our study. This approach suggests that country-specific institutions, which are themselves the products of historical circumstances and social characteristics, influence policy-making.

For example, we might postulate that the differences between the Belgian and Swiss consociational systems versus the U.K. and Canada's Westminster system will lead to differences in their regulation of GMO and ART. Thus, according to the seminal work of Richardson (1982) and Vogel (1986), one might expect that there is a prevailing "national style" of policy-making, which is influenced by both institutional and cultural factors. Comparative politics' traditional focus on political parties and institutions will prove useful in identifying variables that fall under this approach (see Schmidt 1996; Lijphart 1999; Castles 2000; Keman 2000).

## Party Politics: Green and Christian/Neo-Conservative Influence (H3)

The "public choice" perspective in political science presupposes that political parties do not only respond reactively to known social demands (as implicitly suggested by the first two hypotheses). Rather, political parties will, on occasion, initiate a public policy in line with their own preferences. More precisely, the parties establish priorities among issues and then formulate public policies with the goal of expanding their electoral base. They thus broaden the number of potential beneficiaries of the newly proposed policies. The public agenda therefore consists of the issues selected by the main parties, which are often revealed during electoral campaigns and in party programs. The literature suggests a number of variations of this model, with distinctions based on whether the competition among parties rests on ideological differences (Downs 1957; Odershook 1986) or on the specific issues associated with each party (Budge and Farlie 1983; Klingemann et al. 1994). Although there can be little doubt that parties matter, the question remains how they matter in the specific case of biotechnology policy.

Political parties enter into electoral competition on the basis of party programs that they have formulated, proposed to the voters and promised to put in place if given the support of the electorate. Klingemann, Hofferbert, and Budge (1994) have elaborated several hypotheses concerning the influence of electoral programs on the outcome of governmental policies. Their *agenda model* suggests that the importance given to various public policies reflects the priority that had been given in the party programs to various fields of interest, independent of who wins or loses the elections. Their *mandate model* is based on the hypothesis that the party that wins the electoral competition is determinative for the setting of governmental priorities and that the programs of the other parties are not taken into consideration. In effect, the party that wins the election and forms the government cannot disappoint the expectations of its supporters, who were constituted on the basis of promises made during the electoral campaign. The party will therefore try to implement as much as possible of its own program, whereas the issues put forward in the programs of the opposition parties are less likely to be found in the priorities of the government's agenda. Finally, the model of *contagion from the opposition* takes into account the propositions formulated in the opposition parties' programs. The majority in power, in a situation of uncertainty towards its reelection, endorses public policies that are broadly popular, in order to reduce potential threats to the renewal of its mandate.

In order to determine the real impact of parties on public policies in the biotechnology sector, two dimensions must be combined. On the one hand, we need to identify the degree of internal cohesion within the governmental majority and its parliamentary base. This first dimension tries to identify the margin for maneuver of the majority party or coalition to impose a specific policy. Furthermore, we want to identify the position of the various parties towards the development and commercialization of medical and agro-food biotechnologies. This analysis requires an examination of the ideological foundations of the support for, or opposition to, ART and GMOs, as well as the relative importance given to these issues by the various parties.

With respect to the second dimension, we can hypothesize that, for the field of biomedicine, Christian and/or neo-conservative parties will be split against secular and/or progressive parties. The religious and neo-conservative parties will most likely try to implement more restrictive policies in order to promote values like the defense of human life, similar to what was observed during the controversial debates on the partial decriminalization of abortion or other ethical questions (such as homosexual marriage and the liberalization of soft drugs). In terms of the agro-food GMO sector, the partisan polarization is separated between, on the one hand, the Green parties that reject GMOs for environmental reasons and, on the other hand, the parties that do not place this issue at the heart of their electoral program.

The research hypothesis on the influence of political parties can be formulated as follow: *if, in the case of ART, a Christian and/or neo-conservative party is in power and controls the cabinet position in charge of biotechnology, or, in the case of GMO, an environmental party, then the content of the policy will be more restrictive.* Conversely, if government power is in the hands of secular and/or progressive parties (for the case of ART), or parties less responsive to environmental issues (for the case of GMO) then the content of the adopted policy will be more favorable to the interests of the target groups.

## Fragmented Governance: Influence of Federal and Parliamentary Structures (H4)

A country's institutions can either concentrate or fragment policy-making among several arenas or venues. The British state and its Westminster system is a typical case of power concentration. With a prime minister controlling the House of Commons, policy decisions can be made in an authoritative fashion by a single person. In contrast, the United States provides an example of fragmentation. In fact, members of Congress enjoy considerable freedom to vote according to their own preferences rather than those of their political party. The president, therefore, has little control over Congress, even when its majority belongs to the president's political party. Presidents thus frequently veto congressional decisions and Congress sometimes overturns a veto with its two-third majority. And all this fragmentation does not account for the power of courts in the American political systems, nor does it account for the fragmentation created by the country's federal system.

In fact, studies of federalism are particularly informative about the effect of fragmentation on public policy (see Chandler 1987; Watts 1999; Braun 2000; Benz 2002). The federal structures are crucial to the study of biotechnology policies insofar as the formal responsibility of public authorities to intervene in related sectors (the environment, research, consumer protection, or health) is divided between the federal and sub-federal governments. When this is the case, the carrying out of a coherent policy requires a coordinated effort on the part of political leaders and the various levels of administration. The increase in the number of decisional arenas, through the various central/federal and decentralized/regional entities, renders the strategic games of the actors, as well as the results of public policy-making, more complex. As emphasized by Keman, it is not the structures themselves, whether they are federalist or unitary, concentrated or diffuse, centralized or decentralized, that explain the direction and effects of a particular policy but rather the opportunities and constraints, created by these structures, for the participants in the debate:

> What needs to be investigated is the extent to which the behavior of the relevant socio-political actors is more or less shaped by these institutional configurations, and, conversely, to what extent different types of political actors and informal modes of decision-making may well account for the cross-national variation in policy performance. (Keman 2000: 215)

Accordingly, studies of federalism and multilevel governance are increasingly taking into consideration the interaction between federalist structures and both the policy actors and other institutional factors in order to explain the differences in the content of the state interventions in various countries (see Lehmbruch 1998; Grande 2002; Montpetit 2002; Braun 2000; Stepan 2001).

What kind of policy effect can we expect from fragmented institutions? A fragmentation of the spheres of responsibility between various authorities, it has been argued, delay, when it does not entirely block, the decision process, because fragmentation creates veto points for the opponents of state intervention. In the area of biotechnology policy, we expect target groups with permissive beliefs to take advantage of fragmentation by vetoing the adoption of restrictive policies (Tsebelis 1995; 2002). However, this is possible only to the extent that the various policy-making arenas fail at coordinating their policy-making activities. Other policy analysts, however, would disagree, arguing instead that fragmentation multiplies the opportunities for the proponents of restrictive policies. Groups that are opposed or concerned about biotechnology, such as environmentalists or women, can use the numerous forums of fragmented systems to open and expand the policy debate (Baumgartner and Jones 1993). Here again, however, coordination intervenes: a low level of coordination, notably in federal systems, could encourage a regulatory competition whereby policy decisions evolve towards permissiveness. When uncoordinated, policy-making arenas can suffer from the pressure of target groups threatening to move where regulations are permissive or simply end cooperation with the arena. When this occurs, policy-makers tend to make permissive decisions. In short, considerable disagreement exists among policy analysts over the precise policy effects of fragmented structures, including federal institutions (Timmermans 2001).

In order to guide our empirical analysis, we suggest the following hypothesis about the policy effect of fragmented institutions: *if groups opposed to biotechnologies are involved in various policy-making arenas, and if those arenas can coordinate their policy-making activities, biotechnology policy decisions should be restrictive towards target groups.* Conversely, if target groups can confine policy-making to a single site of power, or if coordination between the various policy-making arenas is weak enough to encourage regulatory competition, then the content of the policy should be more permissive.

**Influence of Administrative Structures (H5)**

The "internal expectation" model (Garraud 1990) gives a decisive weight to the administrative actors and public authorities during the drafting of the political agenda and the design of state intervention. These actors, who have already implemented prior public policies, would be in the best position to be able to identify any discrepancies between the state's existing program and the unresolved social problems. Evaluation reports, carried out by the administration, on the effects of a particular policy are often essential pieces of information for anticipating potential future or currently unresolved problems. Administrative actors, taking their own initiative, subsequently propose modifications to past public policies and/or new strategies for intervention. For the proponents of this model, the internal dynamic of the political administrative subsystem is reinforced particularly in cases where a social problem is poorly articulated in civil society. As we saw above, this is often true when issue networks are predominant. In this situation, the public actors take the place of private actors in order to appropriate and redefine the public issue to be resolved. In this way, they guarantee themselves a new legitimacy as a useful public organization, as well as maintain or expand their field of responsibility.

The state is clearly not a monolithic, uniform actor (Carpenter 2004). Much to the contrary, the various administrative departments are defending the interests of the target groups and final beneficiaries of specific public policies, which may often be in competition. For example, in the case of GMO, the departments concerned would likely include the economy, research and agriculture departments, and in terms of the final beneficiaries, the public health and environment departments.

In addition to the fact that each administrative organization pursues the objectives set by its legal and statutory prerogatives, each organization is also motivated by strategies for its own further development, such as the growth of its potential mobilization resources and the reinforcement of its legitimacy as a public institution. Using a "culturist" approach, the sociology of organizations (Crozier 1963) has clearly shown that administrative services do not only consist of formal rules and procedures, but also and primarily embody a system of values, symbols, cognitive frameworks and specific norms of behavior. Administrative institutions are essentially cultural and provide the members of a specific organization with a frame of understanding that guides its members' individual actions.

Taking into account the weight of the organizational "habitus," we can postulate that the different administrative departments will tend to perceive and define the issues facing biotechnology policies in contrasting ways and, thus, propose alternative public programs. We can therefore hypothesize that the process of formulating a public policy and its substantive content derives from the type of administration that has appropriated the responsibility for the topic. This hypothesis has been confirmed in an exemplary manner in the case of biomedicine and agro-food GMO policies in the United States. In fact, Sheingate (2004) explained how the contrasting outcomes of a liberal GMO policy and a restrictive biomedical policy are the result of a conflict between various American administrative agencies. In the GMO field, for example, the Food and Drug Administration (FDA) took a clear product-based approach, while the Environmental Protection Agency (EPA) advocated a process-based approach. The FDA was finally able to impose its views, with the strong support of the White House, at the expense of the EPA's regulatory proposals.

A similar conclusion was reached by the empirical analysis of the regulation of assisted reproductive technologies in Switzerland and Germany. Christine Rothmayr and her colleagues (Rothmayr and Serdült 2004; Rothmayr and Ramjoué 2004) revealed that the very restrictive policies of these two countries were due in large part to the fact that it was the Ministry of Justice that exerted the greatest control over the legislative project, and also tied the development of biomedicine to the legal status of the embryo and the decriminalization of abortion. In these two cases, the administration adopted the role of a veritable policy entrepreneur whose interpretive activities and proposed solutions ended up limiting the range of possible public regulations.

Our fifth hypothesis is as follows: *if the administrative department that takes responsibility for biotech issues tends to represent the interests of the target groups (for example, the departments of agriculture, economy and research for GMOs and public health for biomedicine), then the content of the policy will be permissive.* Conversely, if the administrative agency handling the sector is more likely to represent the views of groups opposing or concerned about biotechnology (for example, the department of the environment for environmental groups in the GMO area), the content of the public policy will be more restrictive towards target groups.

## INTERNATIONALIZATION APPROACH

The last set of working hypotheses focuses on the international factors that might influence policy-making in several countries. In this third category of explanatory variables for biotechnology policies, we examine the formal and informal agreements negotiated at an international level on a specific subject, for example, international commerce, the protection of the oceans, or the problem of global warming. Internationalization is thus to be understood as a collection of principles, conventions, norms and/or decisional procedures that establish the rules of the game, which, in a global context, frame the coordination between national policies and the negotiation between states. Kratochwil and Ruggie (1986: 759) define international regimes, a concept closely associated with internationalization, in the following general manner: "Regimes are broadly defined as governing arrangements constructed by states to coordinate their expectations and organize aspects of international behavior in various issue-areas. They thus comprise a normative element, state practice, and organizational roles."

In the biotechnology sectors, internationalization has steadily progressed since the 1980s. In chapter 2 of this book, we list the principal sources of international norms related to biomedicine and agro-food GMOs (WTO, UN, UNESCO, EU, Council of Europe, FAO or WHO; see table 1.2). The

**Table 1.2.  Overview of International Regimes in the ART and GMO Sectors**

| Regime | | GMO | ART |
|---|---|---|---|
| International level | Sector-specific | - UN Convention on Biological Diversity (1992) and Biosafety Protocol (2000) (Cartagena 2000) <br> - WTO agreements (GATT, TBT, SPS) <br> - Codex Alimentarius of the WHO and FAO | - UNESCO Declaration on the Human Genome and Human Rights (1997) |
| | Cross-sectoral | - WTO agreement about trade-related aspects of international property rights (TRIPS, 1994) | |
| European level | Sector-specific | - Council Directives on contained use of GM Microorganisms (90/219/EEC, revised 98/81/EC) <br> - Council Directive on deliberate release (90/220/EC, revised 2001/18/EC) | - Convention on Human Rights and Biomedicine of the Council of Europe (1997) and Additional Protocol on the Prohibition of Reproductive Cloning (1998) <br> - EP/Council Directive on in vitro Diagnostics (98/79/EC) |
| | Cross-sectoral | - EPA Patent Agreement (EPA) <br> - EP/Council Directive on Legal Protection of Biotechnical Inventions (98/44/EC) | |

influence of these international regulations on the policy design of the nine countries we compare is not obvious for several reasons. We analyzed EU member states and non-European countries. The hierarchy between existing international regulations is not clear and is not uncontested. National authorities have some discretionary power when implementing these international regulations and individual actors strategically mobilize those rules that support their own policy position.

In the following section, we examine the extent to which internationalization, or international rules related to biotechnology, have influenced decision-making processes and the substantive content of domestic biotechnology policies. We focus on two phenomena in particular, which strike us as the most important for our empirical cases: the use of international biotechnology rules by domestic actors and policy transfers.

## Use of International Rules: Changing Opportunity Structures (H6)

The concept of the "Europeanization of public policies" is appearing with increasing frequency in the public policy literature (see Cowles et al. 2001). The nearly exponential growth in articles and books on this topic should not conceal the polysemy of this concept or the diversity of the empirical realities that it covers (Montpetit 2000). According to analysts, the Europeanization of public policies deals with the influence of the member states on the formulation and conduct of European policies (a "bottom-up" approach), the influence of the directives and other regulatory decisions of the European Union on national public policies (a "top-down" approach), and, finally, also deals with the combination of these two processes, which often take place in parallel.

For our empirical analysis, we define Europeanization simply as "a process of change in national institutional and policy practices that can be attributed to European integration" (Hix and Geotz 2000, quoted by Vink, 2003; 64–65). Thus, Europeanization represents a factor that explains institutional transformations (such as administrative reform, or changes in the party system, etc.) as well as changes in the substantive public policies. We have limited our analysis in this context to this second aspect of Europeanization.

Among the authors who have tried to categorize the different types of Europeanization processes on national policies, Knill and Lehmkuhl (2002b) distinguish between three types of mechanisms, which can be categorized from the most prescriptive to the least constraining. The first type of mechanism, "institutional compliance," is due to the fact that the European

Union prescribes a specific model of public policy, which must be integrated as is into the national system of regulations. Under this mechanism, the room for discretionary maneuvering is strictly limited as the new European policy is meant purely and simply to replace existing national policies. The second type of mechanism describes a more flexible process of changing domestic opportunity structures, which consists of changing the rules of the game at the national level and, notably, redistributing resources and altering the power relations between domestic actors. From this second point of view, Europeanization has an indirect effect on the content of policies because some of the parties actively involved in the policy process in question will see their respective influence and interests reinforced to the detriment of the previously dominant actors of the national policy. Finally, the third mechanism deals with the process of "framing domestic beliefs and expectations" that results from the Europeanization of a national policy, which consists of a modification of the cognitive frame of reference created to mobilize support for the new policy. Following the adoption of European norms, the sociopolitical and administrative actors thus modify their representation of the collective problem to be resolved and the possible concrete solutions. This broadening of the debated ideas, values and symbols surrounding the issues at stake in a particular policy produces, as the cognitive approach to public policy informs us, important changes in the implemented national policy.

Two remarks are required at this point. First of all, the three modes of Europeanization identified by Knill and Lehmkuhl are not necessarily mutually exclusive; we can easily imagine, for example, that the prescriptive imposition of a European model results in both the modification of the power relations between local actors as well as the beliefs of these actors.

In addition, these three mechanisms are, more importantly, not necessarily linked to the actions of the European Union. Broadly speaking, they can also be the result of other international factors, such as the policies carried out at the supra- or international levels (for example, the WTO) or be the result of external "focusing events" (Birkland 1998) that affect several countries at the same time (for example, an important scientific innovation). As emphasized by Howlett and Ramesh (2002), it is necessary to analyze the effects of internationalization on public policies by observing how new actors and new ideas influence the preexisting network of public decision-making and then cause a modification of the goals, instruments and/or forms of implementation of the public policy. In any case, it is important to identify correctly the role of the various actors that raise the external factors related to the issue in the media at the national and/or sectoral level and thus legitimize the adaptation of the domestic policy. Public policies are not homeostatic systems that automatically adapt to changes in the international context. To the contrary, we find within the constellation of public policy actors a group that uses the new international events and/or rules to promote their own point of view. We interpret thus the external factors, of which the European policies are an essential component, as real levers of political action that trigger their effects in a particular direction only in relation to the type of actors that effectively make use of them.

Recast in operational terms, the research hypothesis can be stated as such: *if groups opposed to biotechnology and/or their representatives within the political and administrative spheres instrumentalize European policies or other international rules, then the content of the policy will be more restrictive for the target groups.* We should note that the phenomenon of the instrumentalization of the international context could take various concrete forms, as exemplified by Knill and Lehmkuhl's (2002) three modes of Europeanization. As a corollary of our hypothesis, we suggest that if the target groups and/or their political and administrative representatives are the ones who use international rules successfully then the content of the domestic policy will reflect their interests more closely.

## Policy Transfer Across Countries: Learning Processes and Competition (H7)

We now consider the question of the competition or conformity between the biotechnology policies of various countries. Several theoretical frameworks identify public policy-making instruments as the primary element in the transfer of public policies or some of their components. According to Dolowitz and Marsh (2000), this kind of transfer mechanism consists of a "process by which knowledge about policies, administrative arrangements, institutions and ideas in one political system (past or present) is used in the development of policies, administrative arrangements, institutions and ideas in another political system."

The generic term, "policy transfer," encapsulates a number of different processes. For example, the "policy learning" hypothesis describes a process by which the members of a policy network evaluate the efficiency of the instruments that they have previously applied. When dissatisfied with a current framework, they are able to adapt their environment by drawing upon the instruments that were previously and successfully deployed in other sectors or in other public organizations (Bennett and Howlett 1992; May 1992; Rose 1993).

In their meta-evaluation of the theories that propose a causal link between, on the one hand, changes in public policies and, on the other hand,
the processes of policy diffusion, Howlett and Ramesh conclude that the
authors that subscribe to this approach, notably Sabatier (1988), Hall
(1993), and Rose (1993), recognize

> the significance of policy instruments in the process of policy learning as each
> has suggested that, for the most part, in normal times policy learning is in ef-
> fect learning about instruments. It would seem appropriate, then, to suggest
> that theories of instrument choice should be expanded to include the insights
> of theories of policy learning. (Howlett and Ramesh 1993: 14–15)

In addition to these processes of emulation, transfer and voluntary dis-
semination of policy instruments, a number of factors, such as the internationalization of social problems, the global spread of markets and the continued evolution of political integration across regions, all push towards a
broadening of the scope of our analysis. Given the recent evolution of the
context of policy-making, the national and infra-national actors involved in
a specific policy field are working in a more restricted spectrum of possible
actions during the design of their policy. According to Hoberg (1991), there
are three possible hypotheses that describe the roots of this process. In the
first case, countries take into consideration the coordination efforts of existing international regimes. The result is a process of coordinated harmonization of national policies, in line with what we have already explicitly
stated in our previous hypothesis on the internationalization and Europeanization of public policies. In the second case, the economic interdependency of various countries constitutes a potential engine for the convergence of policies among countries, because the strategic program of one
country can cause direct or indirect negative externalities in another country. These externalities imply costs that pressure the country facing these
additional costs to adopt reactively similar instruments to those that have already been implemented by its neighbor. If a country adopts a very restrictive policy in terms of biomedicine or agro-food GMOs, then the national
researchers and manufacturers will switch their activities to the more permissive country. This kind of competition between countries leads eventually to an international race towards the most permissive regulation (a
"race-to-the-bottom") in order to maintain the capacity for scientific innovation and/or commercial production. Finally, the term "penetration" describes the active participation of foreign actors in the formulation of national policies, for example multinational corporations that exert pressure in
order to ensure a uniformity of the national legal frameworks that regulate
its trade in goods and services.

These various theoretical arguments stress the functional interdependency between countries, as well as the strategic aspects to the dispersion of experiences and/or the pressures on national authorities in the context of global competition in the biotechnology sectors. Once again, we need to identify which actors are responsible for the transfer of public policies and the learning processes that support this transfer or argue in favor of a weak public policy for reasons of international competition.

We suggest the following hypotheses for the cases of biomedicine and agro-food GMOs: *if the target groups and/or their political and administrative supporters are able to demonstrate the effectiveness of permissive solutions adopted in other countries, as well as the pernicious effects of more restrictive national laws, then the content of the policy will more likely take into consideration their interests, in particular their economic interests.* Conversely, if the groups concerned or opposed to biotechnology and/or their political and administrative supporters make credible a more restrictive public intervention by presenting a similar policy, which has been adopted in other countries, and that is economically and politically viable, then the target groups will find their activities more restricted by the content of the resulting public policy.

## CONCLUSION

As a synthesis of the previous discussion of independent variables, table 1.3 below presents an overview of our theoretical framework.

This book first and foremost rests on case studies conducted by different country experts. Every country included in this book is the subject of a separate chapter providing a comparison of ART and GMO policy designs (chapters 3 to 11). Each country's chapter applies the theoretical framework presented above in explanations of ART and GMO policy designs. The country chapters are preceded by an overview of international and European Union norms and policies (chapter 2). This overview is a preliminary

Table 1.3.  **Multiple-Comparison Strategy**

| Theoretical approach | Predictions as to variations | Predictions as to commonalities |
|---|---|---|
| Policy Network | Variation across sectors | Commonalities across nations |
| Country Patterns | Variations across nations | Commonalities across sectors |
| Internationalization | Variation across policy designs regulated by different international regimes | Commonalities across policy designs regulated by the same international regime |

Source: Adapted from Levi-Faur 2004: 179

step for understanding the extent to which international rules and domestic factors interact in producing policy designs in individual countries. It should be underlined that the chapter dealing with international and European biotechnology policies is the only one that does not strictly adopt the analytical framework presented in this chapter. As it was conceptualized for the analysis of individual countries, the framework would have required significant adjustments for examining international and European policies. Lastly, we provide in the concluding chapter of the book a full comparative analysis, embracing country and sector comparisons. This last chapter draws upon Ragin's (1987) Comparative Qualitative Analysis, which is a method particularly well suited to account for complex interactions among factors in policy explanations.

We suggest, generally speaking, that policies vary not only considerably across countries, but also across sectors. Some countries have adopted very restrictive policies for both sectors, while others have preferred permissive designs, and some countries have made contrasting choices, adopting restrictions in one sector and a permissive attitude in the other. The individual country chapters provide in-depth sector-oriented analyses of restrictive, intermediary or permissive biotechnology policy designs. The concluding chapter, while underlining some idiosyncratic explanations, focuses on policy-making patterns observable across countries.

## NOTES

1. This limitation to GMO in the agro-food sector and GMO entering the human food chain means that a number of other applications are excluded from our comparative study. We are not looking at the use of GMO in the pharmaceutical industry (e.g., vaccines, medication, transgenic animals for producing medication, and gene therapy, which in some countries is treated like a deliberative release and in others like contained use) and also exclude some uses in the agricultural sector for the moment, namely GM feed, the use of GMO in fertilizer and pesticides, and transgenic farm animals.

2. Since the 1990s the Eurobarometer regularly includes questions on biotechnology. In 1998 the USA, Canada, and Switzerland collaborated on the Eurobarometer on biotechnologies, and Switzerland participated again in 2001.

3. Because not all of the participating countries cover 100 percent of the activities in their country, data are only comparable for about half of the participating countries. For 1999 comparable data was available for Denmark, Finland, France, Iceland, the Netherlands, Norway, Sweden, and Switzerland.

# REFERENCES

Blank, R. H. and S. M. Hines. 2001. *Biology and Political Science*. New York: Routledge.

Bauer, M. W., and G. Gaskell (eds.). 2002. *Biotechnology: The Making of a Global Controversy*. Cambridge: Cambridge University Press.

Baumgartner, F,. and B. Jones. 1993. *Agendas and Instability in American Politics*. Chicago: University of Chicago Press.

Bennett, C. J., and M. Howlett. 1992. 'The lessons of learning: Reconciling theories of policy learning and policy change,' *Policy Sciences* 25 (3): 275–94.

Benz, A. 2002. Themen, Probleme und Perspektiven der vergleichenden Föderalismusforschung, *PVS-Sonderheft* 32, *Föderalismus. Analysen in entwicklungsgeschichtlicher und vergleichender Perspektive*: 9–50.

Bernauer, T., and E. Meins. 2003. Technological Revolution Meets Policy and the Market: Explaining Cross-National Differences in Agricultural Biotechnology Regulation. *European Journal of Political Research* 42 (5): 643–83.

Birkland, T. A. 1998. Focusing Events, Mobilization and Agenda Setting, *Journal of Public Policy* 18 (1): 53–74.

Bleiklie, I., M. Goggin, and C. Rothmayr (eds.). 2004. *Governing Assisted Reproductive Technology: A Cross-Country Comparison*. London: Routledge.

Bobrow, D. B., and J. S. Dryzek. 1987. *Policy Analysis by Design*. Pittsburgh: University of Pittsburgh Press.

Braun, D. (ed.). 2000. *Public Policy and Federalism*. Aldershot: Ashgate.

Bressers, H. Th. A., and L. J. O'Toole Jr. 1998. The Selection of Policy Instruments: a Network-based Perspective. *Journal of Public Policy* 18 (3): 213–39.

Budge, J., and D. Farlie. 1983. Party competition: Selective emphasis or direct confrontation? An alternative view with data. In *Western European Party Systems*. Eds. H. Daadler and P. Mairs. Pp. 267–305. Beverly Hills: Sage Publications.

Carpenter, D. P. 2004. Protection without Capture: Product Approval by a Politically Responsive, Learning Regulator. *American Political Science Review* 98 (4): 613–31.

Castles, F. G. 2000. Federalism, fiscal decentralization and economic performance. In *Federalism and Political Performance*. Ed. Ute Wachendorfer-Schmidt. Pp. 177–95. London: Routledge.

Chandler, W. M. 1987. Federalism and Political Parties. In *Federalism and the Role of the State*. Eds. Herman Bakvis and W. M. Chandler. Toronto: University of Toronto Press.

Cobb, R. W., and C. D. Elder. 1972. *Participation in American Politics: The Dynamics of Agenda-Building*. Baltimore: Johns Hopkins University Press.

———. 1983. *The Political Uses of Symbols*. Longman: New York.

Cobb, R. W., J.-K. Ross, and M.-H. Ross. 1976. Agenda building as a comparative political process. *American Political Science Review* 70 (1): 126–38.

Coleman, W. D. 1998. From Protected Development to Market Liberalism: Paradigm Change in Agriculture. *Journal of European Public Policy* 5 (4): 632–51.

Commission of the European Communities (2002). Life Sciences and Biotechnology: A Strategy for Europe, Brussels: COM(2002) 27 final.

Cowles, M. G., J. Caporaso, and T. Risse (eds). 2001. *Transforming Europe: Europeanization and Domestic Change*. Ithaca and London: Cornell University Press.

Crozier, M. 1963. *Le phénomène bureaucratique*. Paris: Seuil.

Dery, D. 1984. *Problem definition in policy analysis*. Lawrence: University of Kansas Press.

Dolowitz, D., and D. Marsh. 2000. Learning from Abroad: The Role of Policy Transfer in Contemporary Policy-Making, *Governance* 13 (1): 5–24.

Downs, A. 1957. *An Economic Theory of Democracy*. New York: Harper & Row.

Dryzek, J. S. 2000. *Deliberative Democracy and Beyond: Liberals, Critics Contestations*. Oxford: Oxford University Press.

Durant, J., M. W. Bauer, and G. Gaskell (eds). 1998. *Biotechnology in the Public Sphere*. London: Science Museum.

Freeman, G. P. 1986. National Styles and Policy Sectors: Explaining Structured Variation, *Journal of Public Policy* 5 (4): 467–96.

Garraud, P. 1990. Politiques nationales: élaboration de l'agenda. *L'Année Sociologique* 40:17–41.

Gaskell, G., and M. W. Bauer (eds). 2001. *Biotechnology 1996–2000: The years of controversy*. London: NMSI Trading Ltd.

Gaskell, G., P. Thompson, and N. Allum. 2002. Worlds apart? Public opinion in Europe and the USA. In *Biotechnology: The Making of a Global Controversy*. Eds. Martin W. Bauer and George Gaskell. Pp. 351–75. Cambridge: Cambridge University Press.

Grande, E. 2002. Parteiensystem und Föderalismus—Institutionelle Strukturmuster und politische Dynamiken im internationalen Vergleich, *PVS-Sonderheft 32, Föderalismus, Analysen in entwicklungsgeschichtlicher und vergleichender Perspektive*: 179–212.

Hall, P. A. 1993. Policy Paradigms, Social Learning and the State: The Case of Economic Policymaking in Britain, *Comparative Politics* 25 (3): 275–97.

Hassenteufel, P. 1997. *Les médecins face à l'Etat. Une comparaison européenne*. Paris: Press de sciences po.

Heidenheimer, A. J., H. Heclo, and C. T. Adams. 1990. *Comparative Public Policy. The Politics of Social Choice in America, Europe, and Japan*. New York: St. Martin's Press.

Hix, S., and K. Goetz. 2000. Introduction: European Integration and National Political Systems, *West European Politics* 23 (4): 1–26.

Hoberg, G. 1991. Sleeping with an elephant. The American influence on Canadian environmental regulation, *Journal of Public Policy* 2 (1): 107–32.

Howlett, M., and M. Ramesh. 1993. Patterns of Policy Instrument Choice: Policy Style, Policy Learning and the Privatization Experience. *Policy Studies Review* 12 (1/2): 3–24.

———. 2002. The Policy Effects of Internationalization: A Subsystem Adjustment Analysis of Policy Change. *Journal of Comparative Public Policy: Research and Practice* 4 (1): 31–50.

International Service for the Acquisition of Agri-biotech Applications (ISAAA). 2003. Global Status of Commercialized Transgenic Crops: 2002: Preview (C. James); ISAAA Briefs No 27, ISAAA, Ithaca, NY, 2002 (downloaded from the web: www.isaaa.org, 18.8.2003).

Keman, H. 2000. Federalism and policy performance: a conceptual and empirical inquiry. In *Federalism and Political Performance*. Ed. Ute Wachendorfer-Schmidt. Pp. 196–227. London: Routledge.

Klingemann, H. D., R. Hofferbert, and I. Budge (eds). 1994. *Parties, policies and democracy*. Boulder: Westview Press.

Knill, C., and D. Lehmkuhl. 2002. The national impact of European Union regulatory policy: three Europeanization mechanism, *European Journal of Political Research* 41 (2): 255–80.

Krasner, S. D. 1983. *International Regimes*, Ithaca: Cornell University Press.

Kratochwil, F. V., and J. G. Ruggie. 1986. International Organization: A State of the Art on an Art of the State. In *International Organizations* 40 (4): 753–75.

Lehmbruch, G. 1998. *Parteienwettbewerb im Bundesstaat. Regelsysteme und Spannungslagen im Institutionengefüge der Bundesrepublik Deutschland*. Opladen, Westdeutscher Verlag.

Levi-Faur, D. 2004. Comparative Research Designs in the Study of Regulation: How to Increase the Number of Cases without Compromising the Strengths of Case-Oriented Analysis. In *The politics of regulation: institutions and regulatory reforms for the age of governance*. Eds. Jacint Jordana and David Levi-Faur. Pp. 177–99. Cheltenham: Edward Elgar.

Lijphart, A. 1999. *Patterns of Democracy. Government Forms and Performance in Thirty-Six Countries*. New Haven: Yale University Press.

Linder, S., and G. B. Peters. 1984. From Social Theory to Policy Design, *Journal of Public Policy* 4 (3): 237–59.

Linder, S., and G. B. Peters. 1989. Instruments of Government. Perceptions and Contexts, *Journal of Public Policy* 9 (1): 35–58.

Marsh, D., and M. Smith. 2000. Understanding policy networks: towards a dialectical approach. *Political Studies* 48 (1): 4–21.

Marsh, D., and R. A. W. Rhodes. 1992. Policy Communities and Issues Network. Beyond Typology. In *Policy Networks in British Government*. Eds. David Marsh and R. A. W. Rhodes. Pp. 249–69. Oxford: Clarendon Press.

May, P. J. 1992. Policy Learning and Failure. *Journal of Public Policy* 12 (4): 331–54.

Montpetit, É. 2000. Europeanization and Domestic Politics: Europe and the Development of a French Environmental Policy for the Agricultural Sector. *Journal of European Public Policy* 7 (4): 576-92.

——. 2002. Policy Networks, Federal Arrangements, and the Development of Environmental Regulations: A Comparison of the Canadian and American Agricultural Sectors. *Governance* 15 (1): 1–20.

Nygren, K. G., and A. N. Andersen. 2002. Assisted reproductive technology in Europe 1999: Results generated from European registers by ESHRE. *Human Reproduction* 17 (2): 3260–74.

Odershook, P. C. 1986. *Game theory and political theory.* Cambridge: Cambridge University Press.

Pierson, P. 2000. Increasing Returns, Path Dependence, and the Study of Politics. *American Political Science Review* 94 (2): 251–67.

——. 2004. *Politics in Time: History, Institutions and Social Analysis.* Princeton, New Jersey: Princeton University Press.

Ragin, C. C. 1987. *The Comparative Method. Moving Beyond Qualitative and Quantitative Strategies.* Berkeley: University of California Press.

Red Herring. 2003: Biotechnology: The hard cell, by Stephan Herrera (downloaded from the web 18.8.2003: www.redherring.com/investor/2003/02/biotech021303.html).

Rhodes, R. A. W. 1997. *Understanding Governance: Policy Networks, Governance, Reflexivity and Accountability.* Buckingham: Open University Press.

Richardson, J. (ed.). 1982. *Policy Styles in Western Europe.* London: George Allen & Unwin.

Rochefort, D. A., and R. W. Cobb. 1994. *The Politics of Problem Definition.* Lawrence: University Press of Kansas.

Rose, R. 1993. *Lesson-drawing in public policy. A guide to learning across time and space.* Chatham: Chatham House Publishers.

Rothmayr, C., and C. Ramjoué. 2004. Germany: ART Policy as Embryo Protection, In *Comparative Biomedical Policy: Governing Assisted Reproductive Technologies.* Eds. I. Bleiklie, M. Goggin and C. Rothmayr. Pp. 174–90. London: Routledge.

Rothmayr, C., and U. Serdült. 2004. Switzerland: Policy design and direct democracy. In *Comparative Biomedical Policy: Governing Assisted Reproductive Technologies.* Eds. I. Bleiklie, M. Goggin, and C. Rothmayr. Pp. 191–208. London: Routledge.

Rothmayr, C., and F. Varone. 2002. Debate: "Biopolitics" and the missing political scientists: Introduction. *Swiss Political Science Review* 8 (3/4): 129–34.

Sabatier, P. A. 1988. An Advocacy Coalition Framework of Policy Change and the Role of Policy-Oriented Learning Therein. *Policy Sciences* 21 (2):129–68.

Sabatier, P. A., and H. C. Jenkins-Smith. 1999. The Advocacy Coalition Framework: An Assessment. In *Theories of the Policy Process.* Ed. P. A. Sabatier. Boulder: Westview Press.

Salter, B. 2001. Who rules? The New Politics of Medical Regulation, *Social Science and Medicine* 52 (6): 871–83.

Scharpf, F. W. 1993. Positive und negative Koordination in Verhandlungssystemen. *PVS-Sonderheft 24, Policy-Analyse. Kritik und Neuorientierung:* 57–83.

———. 1997. *Games Real Actors Play: Actor-Centered Institutionalism in Policy Research*. Boulder: Westview Press.

Schmidt, M. G. 1996. When parties matter: a review of the possibilities and limits of partisan influence on public policy. *European Journal of Political Research* 30: 155–83.

Schneider, A. L., and H. M. Ingram. 1997. *Policy Design for Democracy*, Lawrence: University of Kansas Press.

Schön, D. A., and M. Rein. 1994. *Frame Reflection: Toward the Resolution of Intractable Policy Controversies*. New York: Basic Books.

Sheingate, A. 2004. *The Politics of Biotechnology in the United States: Medical and Agricultural Applications Compared*. Paper presented at the annual meeting of the The American Political Science Association. Chicago, IL.

Stepan, A. 2001. Toward a New Comparative Politics of Federalism, (Multi)Nationalism, and Democracy: Beyond Rikerian Federalism. In *Arguing Comparative Politics*. Ed. Alfred Stepan. Pp. 315–61. Oxford: Oxford University Press.

Streeck, W., and P. C. Schmitter. 1985. *Private Interest Government*. London: Sage.

Timmermans, A. 2001. Arenas as Institutional Sites for Policymaking: Patterns and Effects in Comparative Perspective, *Journal of Comparative Policy Analysis* 3: 311–37.

Tsebelis, G. 1995. Decision Making in Political Systems: Veto Players in Presidentialism, Parliamentarism, Multicameralism and Multipartyism, *British Journal of Political Science* 25 (3): 289–325.

———. 2002. *Veto Players: How Political Institutions Work*. New York: Russell Sage Foundation.

Van Beuzekom, B. 2001. *Biotechnology Statistics in OECD Member Countries: Compendium of Existing National Statistics*. OECD, Directorate for Science Technology and Industry, STI Working Papers, DTSI/DOC 2001/6 (downloaded from the web: www.olis.oecd.org/olis/2001doc.nsf/LinkTo/DSTI-DOC (2001)6, 18-9-2003).

Varone, F., C. Rothmayr, and É. Montpetit. 2006. Regulating Biomedicine in Europe and North America: A Qualitative Comparative Analysis. *European Journal of Political Research* 45 (2): 317–43.

Vink, M. 2003. What is Europeanization? And other questions on a new research agenda. *European Political Science* 3 (1): 63–74.

Vogel, D. 1986. *National Styles of Regulation. Environmental Policy in Great Britain and the United States*. Ithaca and London: Cornell University Press.

Watts, R. L. 1999. *Comparing Federal Systems*. 2nd ed. Kingston: Institute of Intergovernmental Relations.

Weber, M. 1965. *Essais sur la théorie de la science*. Paris: Plon.

# 2

# TRADE AND HUMAN RIGHTS: INTER- AND SUPRANATIONAL REGULATION OF ART AND GM FOOD

*Gabriele Abels (Bielefeld University)*

In response to the rapid technological and commercial developments in human and nonhuman applications of biotechnology, inter- and supranational regimes started in the 1990s to adopt special policies and to extend existing rules. Global trade, public health, the environment, and even bioethical norms took on a new prominence on the agenda of international regimes.[1] International organizations are essential parts of regimes. The World Trade Organization (WTO), for instance, is considered the core of the global trade regime. From the perspective of international relations, the European Union (EU) is often considered a regime. Yet, it goes beyond traditional regimes given its institutional "thickness," the degree of harmonization of many policy areas, and the status of European law (with direct effect and supremacy over national law) in contrast to international law.

This chapter gives an overview of biopolicies in the agro-food sector (GMOs) and in the field of artificial reproductive technologies (ART) in the context of inter- and supranational regimes. Whereas a number of inter- and supranational regulations dealing with GMOs have come into force, ART issues have been little regulated despite being hotly debated. I will therefore try to account for the differences between the two fields. Given the conceptual framework of this volume (cf. chapter 1), the focus of my analysis is on policy outcomes (i.e., goals, instruments, the implementers, the target groups, and beneficiaries). A comparison between the GMO and ART policy fields highlights interesting differences. I argue that a crucial reason for the stark contrast is the issue of policy framing, which has consequences as to which regime attends to regulation and can lead to potentially competing regimes at the international level. For example, agro-food biotechnology is framed first and foremost as a free trade issue, but ART is, in contrast, primarily framed as a human rights issue on the international level and also within a medical or economic frame on the national level. While both policy frames are deeply rooted in inter- and supranational law, the respective regimes differ. Organizations and instruments for the protection of human rights are generally much weaker insofar as they mainly depend on the political will of the member states or parties to the agreement; trade issues are strongly institutionalized especially in the European Union as well as in the WTO.

## INTER- AND SUPRANATIONAL ART REGULATION

Since the 1980s ART has become widespread. Although ART was already covered by international rules concerning human experimentation (based on the Nuremberg Code and Helsinki Declaration), it was not until the early to mid-1990s that international regimes began to attend to more specific ART and biomedical issues. According to Braun (2000), the evolving regulatory structures instigated a redefinition of the meaning of human rights and dignity. Table 2.1 gives an overview of existing policies. There are no clear inter- or supranational ART regulations as such, but a number of regulations of the United Nations (UN), the Council of Europe (CoE), and the EU affect some aspects of ART.[2] Most regulations, however, establish only "soft laws" (i.e., they establish minimum standards and guidelines, which are binding only to the parties of the respective sections of international law). The choice of concrete policy instruments and enforcement mechanisms is left, by and large, to individual states.

Reproductive cloning of humans is the only prohibited technique and the prohibition consists primarily of nonbinding rules. Several applications of ART, such as the production of embryos solely for research purposes, or sex selection for nonhealth related reasons, are considered problematic and thus constrain the autonomy of researchers and, in some instances, patients. In terms of genetic testing, patients often have free access, thus giving them an overall high degree of autonomy. The most common ART practices such as IVF or preimplantation genetic diagnosis are not covered. The role of the embryo is essential to the debate because it is perceived as particularly vulnerable, but also—so I assume—because embryo research is becoming an increasingly competitive field with economic potential.

**Table 2.1. Overview of Existing ART Policies**

| Phases | Authoritative Decision | Goals | Instruments | Implementers | Target groups | Beneficiaries | Classification |
|---|---|---|---|---|---|---|---|
| Research on human genome [«LABORATORY»] | UNESCO Universal Declaration on the Human Genome and Human Rights | protect the human genome as a common good; protect human rights in relation to genetic knowledge; freedom of science | guidelines | UN Member States | UN member states, implicitly researchers, physicians, laboratories, counselors | everybody | intermediate |
| Human cloning, genetic and embryo research [«LABORATORY and MARKET»] | UN Declaration on Human Cloning | protection of human rights and dignity, protection of the embryo | ban on human cloning | UN Member States that are party to the Declaration | UN member states, implicitly researchers | "embryo" donors | restrictive |
| | Council of Europe (CoE) Convention on Human Rights and Biomedicine plus additional protocol | protection of human rights and dignity with regard to biomedicine | guidelines, prohibition of certain experiments | CoE Member States that are party to Convention | CoE member states, implicitly physicians, counselors, researchers | everybody, "embryo" | intermediate |
| | EU Charter on Fundamental Rights | protection of fundamental rights | guidelines, experimentation ban | EU institutions, EU Member States | EU member states, implicitly researchers | everybody | restrictive |
| | EU Directive 2004/23/EC on Human Tissues and Cells | ensure high protection of human health | quality and safety standards, authorization and licensing, inspection, traceability, notification, guidelines, penalties | EU Member States | manufacturers, physicians | patients, donors | intermediate |
| | EU Commission's Guidelines for R&D | protection of human rights and human dignity in EU-funded R&D in the field of human genetics, embryo and stem cell research | authorization, information and counseling, prohibition of certain experiments | EU Commission | researchers | patients | intermediate |
| Genetic tests and medical devices [«MARKET»] | Council of Europe (CoE) No. R (90) 13 on prenatal diagnosis | ensure freedom of choice and free access, high quality standards for tests and counseling services | guidelines, quality standards | CoE Member States | CoE member states, implicitly physicians, laboratories, counselors | patients | intermediate |
| | Council of Europe (CoE) No. R (92) 3 on genetic testing | fight danger of genetic discrimination and stigmatization by ensuring respect for certain (ethical) principles | guidelines | CoE Member States | CoE member states, implicitly physicians, laboratories, counselors, insurers | everybody esp. patients | intermediate |
| | EU Directive 98/79/EC on in vitro diagnostic medical devices | internal market; ensure safety standards; protect health of patients, users, and third parties | technical standards, registration, labeling, monitoring, penalties | Member States; Commission | manufacturers | patients, users | intermediate |

As described in the overview, the guidelines mainly focus on research; very few guidelines address the market stage. As these policy designs lie in the realm of international law, the regime's member states are thus called upon to implement national legislation in line with international regulation. Given this focus on member states, researchers, laboratories providing services, physicians and counselors are only implicit target groups—except in EU regulations that directly target third parties. The primary goal of these regulations (i.e., the protection of human rights and dignity) is very broad and open to interpretation. Consequently, the final beneficiaries remain vague, with patients being the primary beneficiary but also the rather abstract concept of "the embryo." Given the gendered nature of human reproduction, it is amazing that so few regulations explicitly mention women and their rights as being potentially affected in one way or another by ART. The kinds of policy instruments are mainly "soft law," namely, guidelines; strong rules only apply when ART applications can be framed in terms of market regulation such as in the case of EU directives. Although enforcement of the EU guidelines for research is fairly easy, since the EU Commission itself implements them, prosecution of violators is only possible under national law. Implementation and enforcement thus depend overall on the regime's member states—even for banned practices.

## United Nations

The UN organization for education, science and culture, UNESCO, became involved in ART issues as early as 1993, when it formed the International Bioethics Committee. In the mid-1990s, UNESCO acted in response to the Human Genome Project (HGP), a massive global undertaking to map and sequence the whole human genome, finalized in 2003, which raised a plethora of fundamental social, ethical, and legal questions (for a brief overview see Tauer 2001: 4). In November 1997, the General Conference of UNESCO adopted a "Universal Declaration on the Human Genome and Human Rights" (cf. Reuter 2003). This declaration deployed a strong notion of human rights by reinforcing the dignity of the individual along with the concept of inherent equal and inalienable rights as confirmed in other UN human rights declarations and conventions (Tauer 2001: 231–32). The Declaration states that "in a symbolic sense, it [the human genome] is the heritage of humanity" (Article 1), and it lays out regulatory policy by defining conditions or prohibitions for certain practices involving the human genome. The goal is to protect human dignity in the light of human genome research and to balance the freedom of people and the freedom of science (Article 12). For example, Article 4 declares that "the human genome in its natural state shall not give rise to financial gains," and Article 11 outlaws "practices which are contrary to human dignity" such as reproductive cloning. The Declaration is not effective in and of itself, but rather depends on the individual member states enacting rules and adopting laws. The UN member states are thus the target group and implementers (Article 22). The implicit target group is researchers whose activities are confined; the beneficiary is not so much a particular social group, but the abstract concept of humankind.

In light of continuing scientific and technical progress in the field of cloning, this issue remained on the agenda. Since 2002, there has been fierce debate on a special United Nations "Declaration on Human Cloning," which was finally adopted by the General Assembly of the UN in March 2005 banning "human cloning" (document A/59/516/Add.1). This declaration calls upon UN member states to "prohibit all forms of human cloning inasmuch as they are incompatible with human dignity and the protection of human life," and thus deliberately avoids clarification if this pertains to so-called therapeutic or research cloning. Many countries that supported embryonic stem cell research, and in some cases also allowed for research cloning, voted against the Declaration (e.g., Belgium, France, Sweden, UK, Netherlands, Japan and Canada), because it could be interpreted as a ban on research cloning. This declaration was strongly supported by Germany and Switzerland as well as by the United States due to its strong antiabortion position supported by influential national interest groups belonging to the religious right.

## Council of Europe

The CoE is a regional organization most active in the field of human rights. Its strongest instrument is the "Convention for the Protection of Human Rights and Fundamental Freedoms" enforced by the European Court of Human Rights. In the early 1990s, the Council began to investigate the field of biomedicine in general (not just genetics) and since then has adopted several authoritative decisions that affect ART. The general aim of these decisions is to harmonize legislation via the institutional mechanisms provided by the CoE such as the European Conference of National Ethics Committees (COMETH).

In 1997, the CoE adopted the "Convention for the protection of human rights and dignity of the human being with regard to the application of biology and medicine" (in short: Convention on Human Rights and Biomedicine;

ETS No. 164/1997), which came into force in 1999. Although several of the European countries included in this book have signed the Convention (France, the Netherlands, Sweden, Switzerland), none of them has yet ratified it (neither has the EU). An additional protocol adopted in 1998 regulates the "Prohibition of cloning human beings" (ETS No. 168/1998). The objective of this protocol is to "protect the dignity and identity of all human beings" (Article 1). The Convention refers to various general statutes of international human rights issued by the UN and the CoE itself. It establishes minimum standards, in particular, banning or restricting the use of certain practices. The major focus is on the individual rights of human beings as objects of research. Once again, the member states are the target groups and implementers, and the protocol also implicitly targets researchers, physicians, and genetic counselors. The Convention also takes up some ART-related issues: it bans sex selection for non-health-related purposes (Article 14), calls for protection of the embryo in research (e.g., prohibiting their creation solely for research purposes [Article 18]).

In addition, the CoE adopted two legally nonbinding recommendations: "Recommendation (90)13 on Prenatal Genetic Screening, Prenatal Genetic Diagnosis and Associated Genetic Counselling" and "Recommendation (92)3 on Genetic Testing and Screening for Health Care Purposes." The objectives of "Recommendation (90)13 on Prenatal Genetic Screening, Prenatal Genetic Diagnosis and Associated Genetic Counselling" are to ensure freedom of choice and free access to genetic tests, set high-quality standards for tests and to provide for counseling services. The Recommendation includes a definition of prenatal screening as tests to identify those at risk of transmitting genetic disorders, premarriage and preconception screening, and prenatal diagnosis during pregnancy. It outlines a number of principles: the availability of nondirective counseling (Principle 1, 4); the restriction of tests to severe disorders (Principle 2); the responsibility of the physician and the quality of the approved laboratories (Principle 3); free and informed consent of both partners to all tests (Principle 6, 7); the woman's right to decide free of discrimination (Principle 9, 10); collection, processing and storage of data for medical purposes only (Principle 11); confidentiality (Principle 12); right of access to personal data (Principle 13); and access to preconception counseling and screening (Principle 14).

"Recommendation (92)3 on Genetic Testing and Screening for Health Care Purposes" addresses issues of human reproduction and health insurance. The main topics are quality of counseling and services, confidentiality of data, and equal access to tests. The Recommendation reaffirms the principles of Recommendation (90)13 and suggests 13 (ethical) principles.

It aims at fighting the "dangers of discrimination and social stigmatization which may result from genetic information" by ensuring "respect for certain principles in the field of genetic testing and screening for health care purposes." It covers genetic tests on people who are at risk of transmitting genetic disorders to their offspring.

## EUROPEAN UNION

Biotechnology regulation is covered by the EU common market mandate that includes related areas such as Research and Development. There is no specific supranational ART regulation, although many member states have developed comprehensive frameworks for ART. However, several pieces of EU legislation are relevant to ART-related research and diagnostic devices.[3] "Directive 98/79/EC on *in vitro* diagnostic medical devices" covers medical devices and accessories manufactured from tissues, cells or substances of human origin, including devices for the purpose of providing information on a congenital abnormality (Article 2b). The policy goals of this directive are first to ensure the free movement of these medical devices in the internal market and second to set safety standards. The directive employs technical norms and rules for manufacturing, registration requirements for manufacturers and devices, labeling requirements, quality and safety monitoring. Member states are free to choose the means of implementation, including choice of penalties applicable to infringements. The directive is targeted at manufacturers while the beneficiaries are the users of such devices and, ultimately, patients. In addition, "Directive 2004/23/EC on setting standards of quality and safety for the donation, procurement, testing, processing, preservation, storage and distribution of human tissues and cells" pertains to fetal tissues and embryonic stem cells plus their derivates. The objective is to "ensure a high level of protection of human health" by setting quality and safety standards (Article 1). This directive employs as policy instruments authorization and licensing schemes, technical standards, inspection and control measures, traceability from the donor to the recipient, notification, guidelines for donor selection and data protection, and penalties. The directive is implemented by the member states. The target groups are researchers and manufacturers; the beneficiaries are patients and donors.

The EU enacted guidelines for its research funding policy after years of controversy during the negotiations of the fifth and sixth research framework programs (FP). The EU established a research moratorium for

EU-funded embryo and embryonic stem cell research that was in force until the EU adopted basic ethical principles pertaining to new embryonic
stem cell lines only for FP6 in December 2003 (cf. Abels 2003).[4] These basic ethical principles aim at the protection of human rights and human dignity in EU-funded projects; they introduce procedural and substantive instruments such as authorization, prohibition, information and counseling.
The EU will under no circumstance fund research in a member state that
prohibits such research under its national laws. Funding is limited to donated IVF embryos (thus excluding research cloning) with the added conditions that donors have given free and informed consent; research has to
be approved by an ethics committee, and a scientific committee has to approve of the research project. Finally, the embryos had to have been "produced" before 27 June 2002 so as to avoid their creation solely for FP6 research. Reproductive cloning or intervention into the human germ line is
banned; research cloning is excluded from funding, yet the derivation of
stem cell lines from supernumerary embryos is eligible. This is clearly an intermediate policy compromise, because it introduces some control mechanisms compared to policies at the national level (cf. EU Commission 2003:
38–47; EU Commission 2005), which may be highly stringent and restrict
human embryonic stem cell research (e.g., Germany), or less restrictive and
allow for derivation (e.g., France, Netherlands), or even permissive and allow for research cloning (Belgium, Sweden, U.K.).

These guidelines build on the "Charter of Fundamental Rights" pro-
claimed by the Nice European Council in December 2000. Article 3 prohibits eugenic practices and reproductive cloning, but does not explicitly
mention embryo research. The Charter was then integrated into the EU
Constitution. Given the Constitution's open future due to the negative ref-
erenda in France and in the Netherlands in spring 2005, the status of the
Charter is still weak.

## INTER- AND SUPRANATIONAL REGULATION OF
## AGRO-FOOD BIOTECHNOLOGY

Food is a well-established topic in international politics. New technologies,
especially biotechnology, have subverted the postwar consensus of mutual
recognition of production standards, and the practices of governments have
profoundly changed (cf. Phillips 2001: 27, 36). As table 2.2 illustrates, international regulation of GMOs in the agro-food sector is a "patchwork"
(Phillips 2001: 42). In the course of the 1990s international regimes started

**Table 2.2. Overview of Existing GMO Policies**

| Phases | Authoritative Decision | Goals | Instruments | Implementers | Target groups | Beneficiaries | Classification |
|---|---|---|---|---|---|---|---|
| Contained use [«Laboratory»] | Council Directive 90/219/EC on contained use | effective functioning of internal market; protect human health and the environment | prior notification, standards, inspections and control | competent authorities EU Member States; Commission | researchers | everybody, environment | intermediate |
| Deliberate release [«Experiment»] | Council Directive 2001/18 on deliberate release (previously 90/220) | effective functioning of internal market; protect human health and the environment from risks associated with GMOs | authorization, standards, labeling | competent authorities EU Member States; Commission | researchers, agro-business | everybody, environment | intermediate |
| Quality and safety of food [«Market»] | Codex Alimentarius | quality and safety of food, fair trade | substantial standards (related to SPS Agreement), guidelines | competent authorities in Member States | Codex member states | agro-business, food, processing industry, consumers | permissive |
|  | WTO SPS Agreement | Safe food for consumers, ensure free and fair trade (no arbitrary discrimination) | standards, labeling, sanctions (linked to Codex Alimentarius) | competent authorities in WTO Member States | WTO member states | agro-business, food processing industry, consumers | permissive |
| Production and distribution of food [«Market»] | WTO TBT Agreement | free trade (no discrimination based on origin or other technical standards of products); protection of consumers etc. | technical standards, labelling, sanctions | competent authorities in WTO Member States | WTO member states | agro-business, food processing industry, distributors, consumers | permissive |
| Transboundary movement of GMOs [«Market»] | Cartagena Biosafety Protocol | free trade and protection of biodiversity and human health | standards, information and reporting (AIA), labeling (equal status to WTO agreements) | competent authorities in Member States | Parties to the Biosafety Protocol | everybody, environment | intermediate |
| Safety of novel food and labeling [«Market»] | Council Regulation No. 258/97 on Novel Food and accompanying decisions on labelling | effective functioning of internal market; protect public health and the environment and respect consumer information | authorization, standards, labeling | directly effective in EU Member States, competent authorities; Commission | agro-business, food processing industry, distributors | consumers | intermediate |
| Production and distribution [«Market»] | European Parliament and Council Food and Feed Regulation | effective functioning of internal market; protection of human and animal health and of the environment, ensure consumer information; lay down procedural rules and labeling provisions | standards for assessment, authorization and monitoring, public register, labeling, penalties | directly effective in EU Member States, competent authorities; Commission; EFSA | agro-business, food processing, industry, distributors | consumers, environment, producers of GM-free food | restrictive |
| Production and distribution [«Market»] | European Parliament and Council Regulation on Traceability and Labelling of Food and Feed | effective functioning of internal market; framework for the traceability throughout production and distribution chain | labeling, monitoring and post-market control, penalties | directly effective in EU Member States, competent authorities; Commission | Agro-business, food processing, industry, distributors | consumers, environment, producers of GM-free food | restrictive |

to attend to the GMO field either by developing special rules (EU, UN) or by extending existing rule to GMOs (WTO, Codex Alimentarius).

Within policy designs for the GMO sector, there are differences in the underlying approaches to risk regulation, including the respective policy goals and instruments. Whereas the WTO and Codex Alimentarius regulation is vertical and product-based, the UN Cartagena Protocol and EU legislation is horizontal, process-based and precautionary (see Patterson and Josling 2002: 3–5). Further complicating the picture, we find competition between regimes such as the WTO vs. the EU, and the hierarchy of international norms is not yet clear (Coleman and Gabler 2002). While

> the rules and jurisprudence of international institutions such as the WTO have shaped EU regulations, focusing these on the use of individualized scientific risk assessment as the basis for biotechnology regulation . . . the EU has simultaneously sought to export its domestic regulatory principles, including its version of the "precautionary principle," to the international arena. (Shaffer and Pollack 2004: 5)

The policy design ranges from permissive (WTO Agreements) to rather restrictive (new EU regulations). Two policy rationales dominate both regimes: free and fair global trade on the one hand and the effective functioning of the EU internal market, along with food safety considerations, on the other. Furthermore, environmental concerns are increasingly coming into play.

The inter- and supranational regimes involved in GMO regulation develop rules for single stages in the production and distribution chain—mainly for the final marketing stages. Technical standards are the dominant policy instrument. The EU has developed the most comprehensive and coherent regulatory framework covering the whole chain from the laboratory to the end product. This framework uses a variety of procedural (e.g., authorization) and substantial (e.g., labeling) policy instruments that establish strong pre- and postmarket control; furthermore, enforcement mechanisms are strong. Economic actors in the agro-food sector are often the beneficiaries but are also sometimes the (implicit) target group; the main beneficiary is the consumer and often the environment.

### World Trade Organization

Food safety standards and other technical barriers to trade are becoming more important (Young 2003). "In an increasingly global market, the advantages of having universally uniform food standards for the protection of consumers are self-evident" (Krenzler and MacGregor 2000: 309). Global standards also protect the rights of producers and distributors of food and enable free trade. The WTO is the most important organization in global GMO regulation.[5] The rules and norms of the WTO pertaining to GMOs are not biotechnology-specific, but rather are based on existing jurisdictions (i.e., agreements) which have been extended to GMOs.

The "Agreement on the Application of Sanitary and Phytosanitary Measures" (SPS Agreement) addresses health and safety measures in food. It is closely linked to otherwise nonbinding international standards such as the Codex Alimentarius. The goals are to ensure food that is safe for human consumption and to guarantee that health and safety regulations do not discriminate against import products. The agreement employs standards and sanctions as policy instruments. While countries are allowed to set up their own safety standards to protect human, animal or plant life or health, regulations have to be based on sufficient scientific evidence and may only be temporary (Krenzler and MacGregor 2000: 309).

The "Agreement on Technical Barriers to Trade" (TBT Agreement) covers processing and production methods. It states that products with the same characteristics—so-called "substantially equivalent" or "like products"—have to be treated equally; discrimination is prohibited. So if GM foods are like conventional foods, the first one cannot be treated differently, however, the question of likeness is contested. This agreement sets technical standards such as labeling requirements to be implemented if a legitimate objective such as protection of human health is affected. Countries may establish additional measures only to protect human, animal, or plant life and health, or to protect the environment. Differential treatment solely based on the fact of different production processes (process-based regulation), such as mandatory labeling of GM products, is considered unlawful discrimination.

Both Agreements take a product-based approach according to which GM products as such do not require special treatment, unless they are substantially different from conventional products.[6] Consequently, both Agreements are permissive.

In contrast to other international organizations, the WTO has a fairly strong enforcement mechanism, the dispute settlement procedure, which is to be invoked whenever countries violate an agreement. This creates pressure on the EU regulatory regime (cf. Skogstad 2001). In fact, the main GMO producer countries (United States, Canada, and Argentina) handed in a formal request for consultation under the dispute settlement procedure arguing that the EU policy of a de facto moratorium and the ban imposed by several EU member states is in breach of the two WTO agreements mentioned. In February 2006, the WTO dispute settlement body finally decided in favor of the plaintiffs.

## Codex Alimentarius

The Codex Alimentarius is a joint regulation of the World Health Organization (WHO) and the UN Food and Agriculture Organization (FAO). It is the most important piece of international regulation for processed food. It establishes nonbinding standards for food quality and safety in relation to production, processing, packaging, and commercialization. Yet, because the Codex "is an important point of reference in international law" (Krenzler and MacGregor 2000: 309)—in that it is acknowledged by the WTO—its indirect effects are strong. The underlying risk concept is narrow and excludes hypothetical risks (as opposed to the precautionary principle); thus, the policy is permissive. The goals are the protection of human health, guaranteeing fair trade, plus coordination of all international public and private food standard works.

In July 2003 the Codex Alimentarius Commission adopted the "Principles and guidelines on foods derived from biotechnology," which invokes the concept of substantial equivalence, such that GMOs shall only be placed on the market if they are as safe as their conventional counterparts. The principal effect of these rules is to allow for the implementation of traceability measures.

## Cartagena Protocol on Biosafety

The Biosafety Protocol is an additional protocol to the "UN Convention on Biological Diversity." Negotiated under the auspices of the UN Environmental Program, it was adopted in 2000. Its main achievement is that it introduces environmental concerns into the international trade regime since the Protocol is *not* subordinate but equal to WTO Agreements. It is "the first multilateral instrument to regulate the use, movement and trade of *viable* genetically modified organisms" (Krenzler and MacGregor 2000: 313; emphasis added). Founded on the precautionary principle, the Protocol tries to balance trade and environmental issues, such as the protection of biodiversity. It allows for restrictions on transboundary trade if there is sound scientific evidence of risks to human health or the environment. In contrast to other international (trade) agreements, risk assessment may take scientific uncertainty into account. This protocol also applies to living GMOs intended for direct use as food or feed or in food processing; yet processed products based on GMOs and products in transit are excluded.

The protocol establishes standards and obligations concerning information and reporting (i.e., the Advanced Informed Agreement [AIA]) procedure for first time transboundary GMO movement, and labeling requirements for GMO elements in commodity shipments intended for human consumption (Phillips 2001: 41). Given the constraints on agro-business, the food-processing industry, and distributors, the policy design is clearly intermediate in terms of its restrictiveness. The Convention came into effect in 2003 after being ratified by the fiftieth state. Japan, Switzerland and all EU member states (except for some of the ten new ones) have signed and ratified the Convention as well as the EU itself, which then transposed it into supranational law (cf. below). While Canada has signed but not yet ratified it, the United States has neither signed nor ratified the Protocol.

## European Union

The EU has actively regulated GMOs since the 1990s. The regime was considered intermediate restrictiveness until 2000 when it became more restrictive, catalysed, but not caused, by the BSE crisis (cf. Shaffer and Pollack 2004; Toke 2004, chap. 5). The first EU directives dealing with GMOs, which took a process-based approach, focused mainly on the research stage: "Council Directive 90/219/EEC on the contained use of genetically modified microorganisms" applied to GMOs for research and industrial purposes, and "Council Directive 90/220/EEC on the deliberate release into the environment of genetically modified organisms." These directives covered seeds, plants and microorganisms that contained or consisted of GMOs (e.g., GM maize), although goods produced from GMOs (e.g., tomato ketchup) were excluded. Both directives sought to protect human health and the environment, and to ensure the functioning of the internal market by establishing safety standards for a case-by-case risk assessment procedure. They introduced an authorization process for working with GMOs in the laboratory and for their deliberate release.

Both directives were amended in the 1990s to adjust to technical progress and in light of political criticism. While the contained use directive was deregulated (e.g., lowering safety requirements for some experiments in 1998, now Directive 98/81/EC), Directive 90/220/EC became more restrictive. This directive left authorization for field experiments to the competent authorities (CA) in the member states, but required notification. Authorization for placement on the market involved a two-step procedure combining national and supranational decision-making via the so-called Article 21 regulatory committee. The directive contained a safeguard clause (Article 16) that allows member states—and some have made wide use of this clause—to adopt temporary protective measures in their territories. The revised "Directive 2001/18/EC on the deliberate release into the environment of genetically modified organisms and repealing Council directive 90/220/EEC" introduced, in addition to existing rules, the precautionary principle, more restrictive and extended standards for labeling, mandatory monitoring of long-term effects, a stricter authorization procedure (e.g., first approvals expire after a ten-year maximum period), as well as the possibility for ethical assessment of GMOs. Finally, the simplified notification procedure for like products was suspended. The directives target first and foremost researchers and, in the case of deliberate release, agro-business. The competent authorities in the EU member states as well as the European Commission are responsible for implementation. These regulations establish horizontal (i.e., process-based) rules according to which a special law is required for GMOs, but which left regulatory gaps regarding GM food.

In response, specific regulations for GM food were established in the mid-1990s. "Regulation (EC) No. 258/97 concerning novel foods and novel food ingredients" first introduced vertical (i.e., sectoral) regulation of GM food in 1997, in addition to the existing process-based rules. In so doing, the EU departed from the path of mutual recognition of food.[7] Moreover, by adopting a regulation, the EU opted for its strongest legal instrument; which, unlike a directive, is directly applicable *without* transposition into national laws. The objective of this regulation is to secure the functioning of the internal market for food and to take public health, the environment and consumer information into account. Interestingly, it defines GM food as "novel food" encompassing all foods or food ingredients that have not been widely used for human consumption within the EU before 1997 (Article 1). GM foods and food ingredients is only one of six categories. All novel foods require safety assessment and must undergo a two-step pre-market authorization procedure (Directive 90/220/EEC served as a model); conditions of use can be specified in the authorizations. "Substantially equivalent" food is authorized according to a simplified notification procedure. On top, the regulation introduces a number of substantial policy instruments such a mandatory labeling requirements.

In response to an ongoing political conflict, this GM food regime substantially changed again in 2003: it became more restrictive and is now integrated into the new system for food safety. The differentiation between

GM food and feed is eliminated and common rules are established.[8] "Regulation (EC) No. 1829/2003 on genetically modified food and feed" and "Regulation (EC) No. 1830/2003 concerning the traceability and labeling of genetically modified organisms and the traceability of food and feed products produced from genetically modified organisms and amending Directive 2001/18/EC" include all food, food additives and feed as well as formerly excluded additives and flavorings that either contain, consist of, or are produced from GMOs. Only enzymes are still not covered. The regulation established a centralized risk assessment procedure involving the new European Food Safety Authority (EFSA) and invoked the precautionary principle, which takes hypothetical risks into consideration, as a guiding line for all food policy.

Regulation (EC) No. 1830/2003 aims at a high level of protection of human life and health, animal health and welfare, environment and consumer interests and simultaneously ensures the functioning of the internal market. The regulation employs procedural rules for assessment, authorization and monitoring of GM food and feed, and provisions for labeling. The threshold for labeling the adventitious or technically unavoidable presence of GMOs in food is only 0.5 percent (Article 47). Nonprepackaged GM products (e.g., tomatoes) have to be labeled as follows: "This product contains genetically modified organisms." Authorizations are renewable for ten-year periods upon application. The simplified notification procedure has been abandoned, and a Community register for all authorized GM food and feed has been introduced. The regulation targets agro-business, food-processing industry and distributors. The beneficiaries are first and foremost consumers and producers of GM-free foodstuffs (and in a broader sense also animals and the environment). Member states may lay down rules on penalties applicable to infringements.

Regulation (EC) No. 1830/2003 partially amended the revised Directive 2001/18/EC on deliberate release insofar as it pertains to GM food and feed. It introduced very strong control mechanisms for production and marketing. The goal of the regulation was to provide a framework for traceability that facilitates accurate labeling, environmental monitoring, and withdrawals of products. Labeling became mandatory above a threshold of 0.9 percent; traceability now covers the process "from farm to fork," including production, distribution, sale and consumption. The regulation introduced additional instruments such as postmarket monitoring and controls by tests, registers, etc. The regulation is to be implemented by the member states as well as the Commission itself, and member states may lay down penalties.

In addition, seeds that can be marketed throughout the EU now had to be included in the "Common Catalogue of Varieties of Agricultural Plant Species" administered by a specialized European agency. GMO seed varieties had to be authorized in accordance with the deliberate release directive. If seed is intended for use in food, it also required authorization in accordance with the GM Food and Feed Regulation and it had to be labeled.

The EU is, furthermore, party to the mentioned international treaties, such as the Cartagena Protocol. In July 2003 it adopted "Regulation (EC) No. 1946/2003 on transboundary movements of genetically modified organisms." This regulation aligns EU law with the provisions of the Biosafety Protocol requiring exporters to obtain the authorization of the importing country and introduce measures to segregate crops.

In sum, the new Regulations highly restrict the autonomy of producers and distributors of GMOs, while the autonomy of consumers is very high (choice). All GM food and feed now has to undergo a centralized EFSA licensing procedure. Insofar as production and marketing of GM food (and feed) is principally allowed, the policy design is intermediate. Consequently, the Council of Ministers, when adopting the new Regulations in July 2003, simultaneously lifted the yearlong de facto moratorium. However, the regime establishes very strict conditions for marketing and, furthermore, approvals were, under the new regulations, once again blocked by the Council of Ministers for a period of time. For the first time after the formal end of the moratorium, the EU finally approved the inscription of seventeen varieties derived from Monsanto's 810 maize into the Common EU Catalogue of Varieties of Agricultural Plant Species in September 2004 and, thereby, allowed the seed to be sold in the common market. With respect to food and feed, the Commission in April 2005 extended the authorization for twenty-six GMOs that have already been on the market since 1998 for another nine years. In sum, this has so far been a de facto restrictive policy design. It is now increasingly used by those policy-makers in favor of GMOs, such as the Commission and some member states, for authorizing growing and commercialization of GMOs again in the common market within the given legal framework in response to the "global discipline" of the WTO regime and to exploit the economic potential of GMOs.

## COMPARATIVE EXPLANATION OF ART AND GMO POLICY DESIGN

This overview illustrates that there are vast regulatory differences *between* as well as *within* the two fields. This section focuses more on the differences between the two fields of ART and GMOs, but also addresses some of the differences within the fields.

First of all, the EU has not only the most stringent—and legally binding—regulation of GMOs, but is also the only organization to develop a comprehensive framework covering *all* stages in the production process. ART, however, is a different story. There is, generally speaking, hardly any international regulation of ART issues. Furthermore, risk regulation is a dominant issue, but not the only issue on the agenda. In terms of ART, the overall *goals* are the protection of human rights and of the individual in light of new biomedical developments as well as the protection of human health by ensuring the safety and quality of health services. In the field of GMOs, the major goal is free trade, as well as food safety, and in some cases environmental concerns. Accordingly, in the field of ART patients are the main *beneficiaries* as well as "the embryo," since much of the regulation deals with embryos as research subjects, while consumers are the ones profiting from GMO policies. GMO regulation aims at securing free trade and food safety while protecting human health, the environment, and the rights of consumers. Correspondingly, consumers and the environment are the main beneficiaries, but also producers and distributors of GMOs insofar as they are protected from discrimination. The fact that most ART regulation thus far focuses mainly on the research stage, while GMO regulation concentrates on the market, is due to the respective technological developments in both sectors: GMOs are already on the market, while for those reproductive technologies that are available there is not really an international market (due to the overall national character of medical services), and those that are, in fact, addressed by inter- and supranational regulation (e.g., embryo research) are still experimental. Thus, the main *target groups*—and also the implementers—are member states or parties to the respective regime who are called upon to install national legislation; yet there are also implicit target groups that differ between the fields: while ART regulation appeals first and foremost to researchers, GMO regulation appeals to producers and distributors.

The main *policy instruments* in the ART field are ethical guidelines, and sometimes also more technical standards when products enter the market; and some practices are banned. In the GMO field there is a set of procedural and substantial rules so that technical standards clearly dominate. While the compliance mechanisms are strong in the GMO sector (e.g., WTO dispute resolution mechanisms or the European Court of Justice), mechanisms are weak in the ART field (moral obligation and self-regulation). Finally, the overall policy design regarding ART is permissive, whereas designs for GMOs range from permissive to restrictive policies. Table 2.3 gives a comparative overview of factors explaining the differences.

**Table 2.3. Factors Explaining Differences in Inter- and Supranational ART and GMO Policy Design**

| | ART field<br>hardly regulated | GMO field<br>highly regulated |
|---|---|---|
| International organizations involved | UN and special organizations, Council of Europe, EU | UN special organizations, WTO, Codex Alimentarius, EU |
| Dominant policy frame | human dignity and rights | free trade, also food safety and sometimes environmental concerns |
| Market | mainly national | highly international |
| Market pressure | low | high |
| Economic actors | national; low degree of organization | transnational; high degree of organization |
| Public pressure | low; public interest groups fairly weak | high; well-organized public interest groups |
| Classification | permissive | from permissive to restrictive |

## Policy Frames and Regimes: Trade and Human Rights

Any explanation has to take into account organizations and conflicts at the inter- and supranational level and how they affect policy outcomes; organizations and institutions are the heart of regimes (in analogy to the national pattern approach tested in the individual country studies; cf. chapter 1). Regime differences explain the dissimilarities in the two sectors. GMO regulation is part of the global trade regime, whereas ART regulation is part of the human rights regime. Although both regimes are well rooted in international law possessing regime-specific sets of principles, norms, rules, and decision-making procedures, the trade regime is much stronger, particularly in terms of enforcement. However, the trade regime is not free of human rights issues insofar as consumer protection is, essentially, founded on human rights concerns and so are the economic rights of producers.

The WTO and the EU are both built on an economic policy rationale: free trade and an unrestricted internal market are the respective overarching goals. According to the Europeanization hypothesis, EU regulation strongly influences policies in the individual EU member states, however, EU regulations are increasingly subject to international law such as WTO Agreements. For example, current EU agro-biotechnology regulation is under pressure to prove that it is in line with international law. "Global disciplines" have shaped EU regulation in the field of agro-biotechnology (e.g., case-by-case risk assessment) in addition to pressure arising from "national fears" (Shaffer and Pollack 2004).

The economic rationale certainly pertains to those ART applications that already can or will be treated in the future as trade goods. For example, the future economic potential of products derived from embryonic stem cell research does influence the controversy over research cloning in the EU as well as in the UN. The conflict over research cloning also illustrates that in the ART field there are limits to a framing of the issue purely in trade terms given that human subjects are involved, which introduces a strong cultural and normative dimension. The conflict over a UN declaration on human cloning is the attempt to define the meaning of human dignity and protection of human life in light of new biomedical technologies. Given the vast differences in terms of historical traditions and cultural as well as religious practices on a global scale, ART issues are inevitably difficult to regulate. This challenge of overcoming cultural differences, along with the close link to national health care systems, explains why ART is essentially regulated at the national level—if at all. This weakness in international regulation is, in fact, typical of the human rights field: "International human rights law can be made effective only if each nation chooses to adopt the rules as part of its own domestic legal system." (Tauer 2001: 233). In the arena of human rights, enforcement depends mainly on voluntary compliance (Tauer 2001: 234).

Social norms are also evident in the GMO field with regards to the status of the environment or the cultural meaning of food, concepts of food safety and risk philosophies invoking calls for strong environmental and consumer protection. GM food symbolizes a clash of food safety and eating cultures: "high-tech food" versus "Frankenstein food" or science-based versus politics-based risk regulation. Scholars such as Vogel (2003) or Jasanoff (2005) argue that the current regulatory approach towards new technologies in the EU represents a more risk-averse style compared to US legislation and that biotechnology is a prime example. These underlying risk philosophies, especially when strategically used, and the political culture of science create tensions and incompatibilities. For example, the Biosafety Protocol helps the EU to shield off complaints by other countries against its own restrictive legal framework, since both regulations acknowledge the precautionary principle in risk assessment and management (Shaffer and Pollack 2004: 43). Therefore, the transatlantic conflict over GMOs—essentially the conflict over the hierarchy of a permissive WTO versus intermediate to restrictive EU norms—is not a traditional trade conflict, precisely because it involves regulatory practices and priorities that are rooted in social values and preferences (Falkner 2001: 150; cf. also Skogstad 2001; Young 2003).

### High versus Low Market Pressure

Differences in market pressure also serve to explain regime response and regulatory differences. In the GMO market the stakes are high. The global market value of biotech crops in 2004 was forecasted at $ 4.7 billion (USD). Yet, the market is split on a global scale. Ninety percent of the global hectarage of GM crops is concentrated in four countries: the United States (59 percent), Argentina (20 percent), Canada (6 percent), and China (5 percent), whereas diffusion in the EU member states is about 1 percent (for recent data see James 2004). A breakdown by companies shows that a few transnational chemical and agricultural companies dominate the world market. Furthermore, trade of GM crops between the trading blocs has decreased. While the EU is mostly GM-free, the market in North and South America has rapidly adopted GMOs. Producers have a massive interest in exporting GM crops to the EU, and they certainly want to avoid cost-intensive provisions they consider discriminatory against GMOs and protectionist. Consequently, the producer countries try to enforce the global trade and food safety regimes and aim to extend the existing regulatory framework of product-based risk regulation to GMOs by employing the WTO dispute settlement procedure.

In the ART field there exists so far at best a small *international* market. Of course, medical devices, diagnostic kits, etc. are traded as goods on a global market, but with regard to ART applications such as IVF services or prenatal testing there is only a very small *international* market, notwithstanding that there is some transborder "reproductive tourism."[9] In most European countries, the ART market is strongly nationalized and regulated due to the highly regulated national health care systems that often cover expenses, unlike the situation the United States where there is a vast ART market due to the privatization of services and the extensively unregulated market (cf. chapter 3). However, there is as yet no international organization that has a regulatory mandate in this field.

### Public Opinion and Public Interest Groups

Public pressure is one of the chief factors that led to an increasingly restrictive legal framework for GMOs in the EU (Bernauer 2003; Bernauer and Meins 2003; Shaffer and Pollack 2004). According to Eurobarometer data, "a majority of Europeans do not support GM foods, which are judged to not be useful and to be too risky to society. For GM crops, support is lukewarm; although GM crops are judged to be moderately useful, they are seen as almost as risky as GM foods" (Gaskell, Allum, and Stares 2003: 1). Between 30 percent and 65 percent of Europeans reject GMOs for one reason or another. Risk is a major concern, but they also see no social need for GMOs, especially GM food. So when US companies tried to export GMOs to the EU in the mid-1990s, this resulted in consumer boycotts, uprooting of field trials and public protests in many EU countries (cf. Bauer and Gaskell 2002). In the course of the GMO struggle, a strong anti-GMO network was formed; public interests groups (e.g., Greenpeace) in the GMO field used critical public opinion as a campaign resource (cf. Toke 2004, 185f.). Consumer groups and environmental interest groups joined forces in a campaign against GMOs and they made use of the critical public opinion as a mobilizing resource. The number of food scandals in the EU in the 1990s (e.g., BSE, dioxin contaminated meat) certainly influenced the GMO debate. European policy-makers had to respond to the massive level of public protest; in effect, the regulatory debate in the EU became closely linked to the general governance debate and GMO became a question of European identity (cf. Abels 2002; Jasanoff 2005: chap. 3).

On the other side of the Atlantic, public attitudes about GMOs were very different and overall supportive (cf. Gaskell, Thompson, and Allum 2002). However, in the early days of the regulatory debate in the United States there was intensive organizational competition over the right regulatory strategy; whereas the Environmental Protection Agency favored a process-based approach, the Food and Drug Administration, which in the end took over the policy field, pushed for product-based regulation (cf. Toke 2004: chap. 4). Today there is, for instance, no special regulation for GMOs, but a product-based approach prevails along with a free market approach. Labeling, for example, is voluntary. There is no dominant strand of critical public opinion and hardly any public interest group is active in trying to influence the current Bush administration and its position on international regulation. Existing international regimes such as the WTO and Codex Alimentarius clearly work in favor of US regulation.

In the ART field, public attitudes are overall much more positive and appreciative of scientific developments. Nonetheless some applications of ART are considered problematic (e.g., use of techniques for nonhealth related purposes or research involving embryos). The regulatory debate focuses primarily on the national level and there has been hardly any pressure for international regulation of ART, the exception being the "UN Declaration on Human Cloning." Essentially, there is no public pressure and no transnational movement or interest group such as medical associations or patient groups pushing for international harmonization or even supranational regulation. Unlike in the GMO field, where the beneficiaries of policies—be it producers, be it consumers—have been strongly organized and highly active in lobbying for regulation, the beneficiaries of ART policies, above all patients, are largely absent in the international arena.

## CONCLUSION

Biotechnology has been on the inter- and especially supranational agenda for more than a decade. The current status is marked by regulatory polarization and differentiation. There are several competing regimes in both fields, but especially regarding GMOs. Policy designs differ immensely and range from permissive, intermediate and restrictive policies. While a strong trade-related policy frame prevails, it is contested because of the "inherently multisectoral nature of GMO regulation" (Shaffer and Pollack 2004: 4) Thus, the heated debate over GM food presents a search for a new consensus that allows for integration of safety and environmental concerns into the narrowly defined, neo-liberal trade regime that strongly relies on "sound science." Since the—economic and cultural—stakes are high, finding solutions is difficult.

In the ART field the regulatory regime is less developed and so far dominated by a human rights framing and by a very limited scope on cloning; this goes along with a lack of strong enforcement rules. Overall, regulations tend to be permissive insofar as most techniques are not regulated; although for human cloning there is a restrictive design. This situation contrasts with the often very comprehensive ART policy designs that we find in many countries. ART applications are more easily regulated internationally when they result in commercial products. This then allows for their integration into the trade regime, but they are often still treated as special goods requiring bioethical safeguards (e.g., human tissues, cells or embryos). Yet given the strong links of most ART applications to national health care systems, an extensive inter- or supranational regulation is most unlikely and will be restricted to a few issues.

There is clearly a strong bioethical dimension prevalent in ART regulation, yet this is not restricted to this field. There is a gradual expansion of bioethical norms to all policy debates over GMOs in the EU such that ethics has taken center stage today (Lindsey et al. 2001). The CoE provides a forum for debate among national ethics commission and, in the long run, may also enable some legal harmonization; the European Commission has its own advisory group, the European Group on Ethics, which closely cooperates with similar bodies in the member states. Nevertheless, this trend is not yet strong at the international level, but may increase in the long run. The development of public opinion—and of consumer behavior—may contribute to change, along with technological development itself. The most important factors are the future of the EU regime for GMOs, the compatibility of the regulations now in force with existing international norms and the status of the precautionary principle.

## NOTES

I am grateful to Daniel Barben for comments on an earlier version of this chapter.

1.  I use the term *regime* here as defined in regime theory by Krasner: "Regimes can be defined as sets of implicit or explicit principles, norms and rules and decision-making procedures around which actors' expectations converge in a given area of international relations. Principles are beliefs of fact, causation and rectitude. Norms are standards of behaviour defined in terms of rights and obligations. Rules are specific prescriptions or proscriptions for action. Decision-making procedures are prevailing practices for making and implementing collective choice." (Quoted in Rosamond 2000: 167)

2.  One of the first international organizations attending to biomedical issues was, in fact, the African Union. It adopted a "Resolution on Bioethics" at its 32nd. Ordinary Session in Yaoundé, Cameroon, in July 1996. It addressed biomedical research as a human rights issue; it referred to the UN Declaration on Human Rights while at the same time appreciating the benefits of scientific progress and stressing the need for international cooperation. As a resolution its legal status is weak.

3.  These devices may also be patentable according to EU Directive 98/44 on the legal protection of biotechnological interventions, which, however, excludes "processes for cloning human beings" as well as "uses of embryos for industrial or commercial purposes." A Directive is a legal instrument that fixes the goals to be achieved but allows the EU member states to choose the best means.

4. The European Parliament has adopted initiative resolutions on artificial in vivo and in vitro insemination as well as on the ethical and legal problems of human genetic engineering as early as 1989 and four resolutions against human cloning between 1993 and 2000. Advisory bodies of the European Unions also produced opinions on embryo and stem cell research in 1998.

5.  In addition, the OECD has assisted in the development of international norms (e.g., by providing information exchange and dialogue).

6.  A third relevant agreement deals with trade-related intellectual property rights (TRIPS).

7.  Foodstuffs commercialized before Regulation 258/97 came into force were still marketed under this principle.

8.  This is in response to the StarLink scandal in the United States, where GMO maize authorized for use in feed only was found in taco chips for human consumption.

9. For example, patients travel to countries with more permissive rules to have access to reproductive services they are not eligible for (e.g., fertility treatment for lesbians or single women in Spain) or to use practices that may be outlawed in their home country (e.g., German or French couples going to Belgium or Italy—at least until restrictive legislation came into effect in 2004—to use techniques outlawed in Germany or France such as preimplantation genetic diagnosis or sex selection). Some couples may simply want to save money and go to countries that provide cheaper ART services or donor gametes such as eggs (for example, US couples going to Italy or buying eggs from Romanian women).

## REFERENCES

Abels, G. 2002. Experts, Citizens, and Eurocrats—Towards a Policy Shift in the Governance of Biopolitics in the EU. *European Integration Online Papers* 6 (19). http://eiop.or.at/eiop/texte/2002-019a.htm.

———. 2003. The European Research Area and the Social Contextualisation of Technological Innovations: The Case of Biotechnology. In *Changing Governance of Research and Technology Policy: The European Research Areas*. Ed. J. Edler, S. Kuhlmann, and M. Behrens. Pp. 314–35. Cheltenham: Edward Elgar.

Bernauer, T. 2003. *Genes, Trade, and Regulation: The Seeds of Conflict in Food Biotechnology*. Oxford: Princeton University Press.

Bernauer, T., and E. Meins. 2003. Technological Revolution Meets Policy and the Market: Explaining Cross-National Differences in Agricultural Biotechnology Regulation. *European Journal of Political Research* 42 (5): 643–83.

Bauer, M. W., and G. Gaskell (eds.). 2002. *Biotechnology: The Making of a Global Controversy*. Cambridge: Cambridge University Press.

Braun, K. 2000. *Menschenwürde und Biomedizin. Zum philosophischen Diskurs der Bioethik*. Frankfurt/M., New York: Campus.

Coleman, William D., and M. Gabler. 2002. Agricultural Biotechnology and Regime Formation: A Constructivist Assessment of the Prospects. *International Studies Quarterly* 46 (4): 481–506.

EU. Commission of the European Communities. 2003. *Report on Human Embryonic Stem Cell Research*. Commission Staff Working Paper SEC 441 (3 April 2003).

———. 2005. "How Does the European Commission Deal with Ethical Issues within its Framework Programme for Research and Development?" Rapid Press Releases MEMO/05/121 (8 April 2005).

Falkner, R. 2001. Genetic Seeds of Discord: The Transatlantic GMO Trade Conflict after the Cartagena Protocol on Biosafety. In *Governing Food: Science, Safety and Trade*. Eds. P. W. B. Phillips and R. Wolfe. Pp. 149–61. Montreal: McGill-Queen's University Press.

Gaskell, G., N. Allum, and S. Stares. 2003. *Europeans and Biotechnology in 2002: Eurobarometer 58.0*. 2nd ed. London.

Gaskell, G., P. Thompson, and N. Allum. 2002. Worlds apart? Public Opinion in Europe and the USA. In *Biotechnology: The Making of a Global Controversy.* Eds. M. W. Bauer and G. Gaskell. Pp. 351–75. Cambridge: Cambridge University Press.

James, C. 2004. Preview: Global Status of Commercialized Biotech/GM Crops: 2004. *ISAAA Briefs* 32. Ithaca, NY: ISAAA.

Jasanoff, S. 2005. *Designs on Nature: Science and Democracy in Europe and the United States.* Princeton, NJ: Princeton University Press.

Krenzler, H. G., and A. MacGregor. 2000. GM Food: The Next Major Transatlantic Trade War? *European Foreign Affairs Review* 5 (3): 287–316.

Lindsey, N., M. Wambui Kamara, J. E. Erling, and A. T. Mortensen. 2001. Changing Frames: the Emergence of Ethics in European Policy on Biotechnology. *notizie di POLITEIA* 17 (63): 80–93.

Patterson, L. A., and T. Josling. 2002. Regulating Biotechnology: Comparing EU and US Approaches. *European Policy Papers* 8. http://aei.pitt.edu/archive/00000028/01/TransatlanticBiotech.pdf (accessed March 7, 2005).

Phillips, P. W. B. 2001. Food Safety, Trade Policy and International Institutions. In *Governing Food: Science Safety and Trade.* Eds. P. W. B. Phillips and R. Wolfe. Pp. 27–78. Montreal: McGill-Queen's University Press.

Reuter, L. 2003. *Modern Biotechnology in Postmodern Times? A Reflection on European Policies and Human Agency.* Dordrecht, Boston, London: Kluwer.

Rosamond, B. 2000. *Theories of European Integration.* Basingstoke, New York: Palgrave.

Shaffer, G. C., and M. A. Pollack. 2004. Regulating Between National Fears and Global Disciplines: Agricultural Biotechnology in the EU. *Jean Monnet Working Paper* 10/04. www.jeanmonnetprogram.org/paper/04/041001.pdf (accessed March 7, 2005).

Skogstad, G. 2001. The WTO and Food Safety Regulatory Policy Innovation in the European Union. *Journal of Common Market Studies* 39 (3): 485–505.

Tauer, J. E. 2001. International Protection of Genetic Information: The Progression of the Human Genome Project and the Current Framework of Human Rights Doctrines. *Denver Journal of International Law & Policy* 29 (3/4): 209–38.

Toke, D. 2004. *The Politics of GM Food: A Comparative Study of the UK, USA and EU.* London, New York: Routledge.

Vogel, D. 2003. The Hare and the Tortoise Revisited: The New Politics of Consumer and Environmental Regulation in Europe. *British Journal of Political Science* 33 (4): 557–80.

Young, A. R. 2003. Political Transfer and "Trading Up"? Transatlantic Trade in Genetically Modified Food and U.S. Politics. *World Politics* 55: 457–84.

# 3

# DIFFERENT PATHS TO THE SAME RESULT: EXPLAINING PERMISSIVE POLICIES IN THE USA

*Francis Garon and Éric Montpetit*
*(Université de Montréal)*

The United States is a country of paradoxes. It is a country with unprecedented contributions to scientific progress and at the same time an astonishing proportion of its citizens reject evolution theory. In the American policy sciences, the theories of rational decision-making encourage a strong belief in a single, correct way to make policies (Stone 1997), and yet Americans scholars are proud of their country's political institutions, which enable the coexistence of several parallel and often conflicting policy-making processes. As this chapter will show, the politics of the knowledge economy in the United States is not deprived of paradoxes and has a unique flavor in comparison to the other countries covered in this book.

The first two sections of this chapter illustrate the extent to which the American ART and GMO policies are permissive. Given the paradoxical nature of the political culture, we would not necessarily expect consistent results. However, we argue that the two policies are permissive for two different reasons: in the ART sector, policy-makers are unable to make decisions, thereby promoting a laissez-faire policy orientation; in the GMO sector, policy-makers deliberately made permissive decisions. Permissive decisions in the GMO sector, we argue in the third section of the chapter, are locked in by a cohesive network of industry actors, as well as by supportive political parties. In contrast, the inability to make decisions in the ART sector stems from the presence of divided networks of actors. Being well connected to decisive state institutions, the actors have the capacity to veto one another's policy initiatives. In other words, policy networks in the ART sector do not provide strong support for permissive policies, but they are also incapable of adopting restrictive policies. Finally, we are unsurprised to note that globalization has had little influence on American biotechnology policy, given the power exerted by the United States on the international scene.

## PERMISSIVE ART POLICY

As table 3.1 indicates, the United States' ART policy is permissive in all important respects. As we have done for other countries, we distinguish between three dimensions of ART policy design: research and experimentation, the practice of fertility treatments and measures to protect patients. The research and experimentation on humans dimension raises concerns that are distinctive from those regarding the practice of fertility treatments and related measures to protect patients. While fertility treatments are generally accepted medical practices in the United States, research, in particular embryo research, is far more controversial. As discussed at length below, embryo research is entangled with abortion, an issue still vigorously debated in the United States. Therefore, while fertility clinics carry out their practices unhampered by policy controversies (US Department of Health and Human Services, CDC 2004: section 5), embryonic research has been the subject of lively debates. Nevertheless, embryonic research has suffered very little restrictions as the result of policy decisions.

The actors involved in the policy debate on embryonic research in the US hold sharply contrasting views, mirroring for the most part the opposition between pro-life and pro-choice actors over abortion. We discuss the nature of this opposition further in the section below on policy networks. The tense nature of the debate has made legislative action particularly difficult. Indeed, none of the forty-five bills introduced in Congress since 1996 to restrict cloning and embryonic research was adopted (Sheingate 2006). This leaves federal funding for research, which is largely but not exclusively controlled by the executive branch, as the only policy instrument to govern embryonic research. Given that private firms and private foundations finance an important percentage of the research conducted in the United States, the efficacy of this policy instrument is limited, as it leaves privately funded laboratories free from policy restrictions. Despite the limitations on the funding instruments, the American embryonic research policy is thus very permissive. It is often argued that the UK's ART policy is more permissive

Table 3.1. Current American ART Policy Design

| | Research | Practice of fertility treatments | Measures aimed at patients |
|---|---|---|---|
| Policy Design | - No national legislation for research conducted with private funds<br>- Since August 2001, public funds provided by NIH under condition<br>- There are still legislative debates on the subject | - Mostly self-regulated (ASRM, SART)<br>- States' historical responsibility to regulate medical practices | - No limitation of access for infertile couples that have the financial resources to pay for treatment |
| Type | Permissive | Permissive | Permissive |

than the American ART policy because the UK's policy was recently changed to permit therapeutic cloning. However, a Massachusetts-based private company, Advanced Cell Technology, announced in *Scientific American* that it had been conducting research on human cloning for several years and had successfully cloned the first human embryo on American soil in October 2001 (Cibelli, Lanza, West, and Ezzell 2002).

The history of restrictions related to the use of federal dollars to fund embryo research began early in the 1970s in response to concerns about the ethics of IVF treatments. In 1973, Congress created the National Commission for the Protection of Human Subjects of Biomedical and Behavioral Research. This body was responsible for, among other things, offering advice on human fetal and embryonic research. One of its recommendations was the creation of a national Ethics Advisory Board within the Department of Health, Education, and Welfare (DHEW, now the Department of Health and Human Services). This Ethics Advisory Board would notably provide guidelines regarding the conduct of embryo and fetal tissue research. The board, based within DHEW, was officially appointed in 1978 and regulations (45DRF46.204[d]) required it to provide advice on publicly funded embryonic research. The board's advice was that IVF research, the main kind of embryonic research existing at the time, was acceptable from an ethical point of view. However, the allocation of funding for research projects fell under the responsibility of the DHEW, which turned out to be an apparently difficult task (Gunning and English 1993). Indeed, the task appeared so difficult that in 1980, when the Ethics Advisory Board's charter expired, the DHEW decided not to renew it. However, the regulations requiring research projects to be examined by the board remained unchanged, creating a de facto ban on publicly funded embryonic and fetal tissue research.

While a number of state governments passed legislations relevant to ART in the 1970s and in the 1980s (Goggin and Orth 2004), the policy situation remained unchanged at the federal level throughout the 1980s. The first federal policy initiative regarding ART research since 1980 arrived in 1993 when Congress adopted the National Institutes of Health (NIH) Revitalization Act. While the act itself did not change the ART policy, it enabled the executive branch to revisit this issue of federal funding for embryonic research. The Clinton administration, however, failed to act quickly enough on this matter. Indeed, before the NIH could spend a single dollar on embryonic research, control of Congress switched to the Republican Party. In 1995, Congress passed the Dickey Amendment, which constrained the NIH's capacity to fund research involving human embryos. Since then, the amendment was reenacted every year (The President's Council on Bioethics 2004: 26) and the Clinton administration simply decided to continue with the ban.

Interestingly enough, the ban on federal funding of embryonic research was partially lifted by the Republican administration of George W. Bush in August 2001. Asserting that his decision was in compliance with the Dickey Amendment, Bush enumerated a set of conditions under which the NIH could fund embryonic stem cell research. These conditions were that: (1) the derivation process be initiated prior to 9:00 p.m. EDT on August 9, 2001; (2) the stem cells be derived from an embryo that was created for reproductive purposes and was no longer needed; (3) informed consent be obtained for the donation of the embryo and that the embryo donation not involve financial inducements. A task force was then created within the NIH with the mandate to "enable and accelerate the pace of stem cell research by identifying rate limiting resources (both material and human) and develop initiatives to enhance these resources; and seek the advice of scientific leaders in stem cell research about the challenges to moving the stem cell research agenda forward and strategies NIH may pursue to overcome these challenges."[1] In the fiscal year 2002, the NIH spent approximately 11 million dollars on human embryonic stem cell research.[2] In addition, the 2001 decision effectively left researchers with only nineteen stem cell lines on which to conduct research from public funds. At the time of writing, the US Senate had yet to take action on the Stem Cell Research Enhancement Act, which would override Bush's funding conditions and increase the number of lines and level of spending on stem cell research. Bush, however, announced he would veto the bill. We discuss this matter further in the last section of this chapter, but may summarize at this point with the observation that the 2001 decision provided access to few lines and little public funding for embryonic stem cell research. The state of California has even had a lively debate on the creation of a state fund to make up for the absence of federal funding for stem cell research. We should recall also that the American policy design leaves private firms and foundations free to spend as much as they want on any stem cell lines, indeed on any type of research involving human embryos and fetal tissues.

The American policy regarding fertility treatment, dealing with the right of couples to have children, is much less controversial and has been largely regulated by state governments, with only New Jersey restricting access. In all other jurisdictions, policies do not limit access based on age, martial status, or the sexual orientation of patients (Goggin and Orth 2004: 83). IVF treatments appeared in the United States in 1981 (U.S. Department of Health and Human Services, CDC 2004: 1) and fertility clinics have been ever since largely self-regulated. Voluntary general guidelines are provided by two organizations: the American Society for Reproductive Medicine (ASRM) and the Society for Assisted Reproductive Technology (SART). Interestingly enough, the 1992 Fertility Clinic Success Rate and Certification Act upheld self-regulation as the key instrument to govern fertility clinics. The act indeed prevents the federal government from adopting "regulation, standard, or requirement which has the effect of exercising supervision or control over the practice of medicine in assisted reproductive technology programs."[3] The Act simply requires American fertility clinics to report pregnancy success rates to the Department of Health and Human Services' Centers for Disease Control and Prevention (CDC). The CDC then annually reports the success rate of clinics for the past seven years. Exemplifying again the importance of self-regulation in this area, clinics submit their data to SART, an association that maintains a list of clinics known to perform ART treatments. In the CDC's 2004 report of success rate, 428 clinics were listed (US Department of Health and Human Services, CDC 2004: 479–510).

As Goggin and Orth (2004: 83) put it,

ART policy design provides a high degree of autonomy to ART researchers and practitioners. In the United States, federal government agencies have been used to protect human subjects and guarantee the autonomy of scientists as long as these researchers are not asking for federal funds to conduct research on fetal tissue or embryonic stem cells, or to experiment with human cloning. Physicians who practice ART are essentially self-regulated, thus ensuring their autonomy, at least in the sense that they can decide what ART to practice.

## PERMISSIVE GMO POLICY

Like the ART policy design, the American GMO policy design is permissive (see table 3.2). Developers of GM food and cultivars are not constrained by intrusive rules and regulations, as a large part of the regulatory process relies on voluntary measures. Moreover, unlike embryo research, GMOs have not been at the center of significant controversies, given that the issue has been framed in scientific and economic terms rather than in ethical terms. Therefore, the American GMO policy remained largely unchanged since its development in the middle of the 1980s.

Discussions on the development of a GMO policy began seriously in 1984, when the White House's Office of Science and Technology Policy (OSTP) proposed a regulatory framework to deal with this emerging issue. After two years of work on the proposal, the OSTP presented in June 1986 the Coordinated Framework for the Regulation of Biotechnology, which remains to this day the key policy document regarding GMOs. The framework adopts a product-based approach, which, in a manner similar to Canada, employs existing statutes and regulatory agencies to govern the products made through rDNA technologies. The agencies granted regulatory responsibilities for GMOs under the framework are: the United States

Table 3.2.   Current American GMO Policy Design

|  | Field Trials | Commercialization and Cultivation | Consumer and Environmental Protections |
|---|---|---|---|
| Policy Design | - Since 1992 persistent deregulation and streamlining of the regulatory system (product-based)<br>- Government permits, delivered by APHIS (USDA), required only for limited products that are listed as plant pest | - Identical approach to that applied to foods developed by traditional plant breeding (FDA)<br>- Voluntary premarket consultation concerning the safety of foods (FDA) | - No specific requirements for labeling and traceability, unless nutritional or health aspects of the food have changed in a substantial way |
| Type | Permissive | Permissive | Permissive |

Department of Agriculture (USDA), more specifically the Animal and Plant Health Inspection Service (APHIS), the Environmental Protection Agency (EPA), and the Food and Drug Administration (FDA). The authority of these three agencies over GMOs rests on the Federal Plant Pest Act (USDA / APHIS), the Plant Quarantine Act (USDA / APHIS), the Federal Food, Drug, and Cosmetic Act (FDA), the Federal Insecticide, Fungicide, and Rodenticide Act (EPA), and the Toxic Substance Control Act (EPA). Congress adopted all of these acts prior to the use of rDNA technologies in the production of cultivars and food. In 1986, very few people in the United States were worried about the environmental and health effects of biotechnology, as the economic potential of biotechnologies was considered far more important. Therefore, the Coordinated Framework's objective was to prevent anything from slowing down the development of the biotechnology industry, while providing policy-makers with enough flexibility to adjust the policy in line with the pace of scientific development (National Research Council 2002: 50).

According to the framework, the USDA and the EPA, depending on the nature of the cultivars, are responsible for overseeing field experiments. The USDA's APHIS decides when GM seeds pose sufficient agronomic and environmental risks to be subjected to regulations and when they are safe enough to be freed from the agency's oversight. When the history of parent plants and of the inserted traits provide sufficient guarantee that a cultivar is safe, it is deemed familiar and will not be treated as a regulated article. In contrast, when a GM seed is deemed unfamiliar, environmental release is decided on the basis of direct experience through field-testing under permit conditions. Any removal of an unfamiliar GM cultivar from the oversight of APHIS involves a petition process whereby an advertisement is published in the US Federal Register and a public comment period is provided (MacKenzie 2000: 50; Vogel 2002: 5). In 1993, the USDA streamlined its process by introducing a notification procedure for tomato, corn, tobacco, soybeans, cotton, and potatoes as an alternative to its permits for field-testing unfamiliar seeds. This notification procedure was extended in 1997 to cover 99 percent of GM plants (National Research Council 2002: 107). Under the notification procedure, the developer simply informs USDA about the sites of testing, imports and interstate movements of its seeds. This streamlined procedure provides a good illustration of the permissiveness of the American GMO policy design.

Another illustration of permissiveness is the limited responsibilities of the EPA for GM cultivars. Indeed, the agency oversees only plants producing their own protection against pests such as Bt maize. Thanks to a genetic alteration, this maize produces a protein capable of destroying the European corn borer, a common pest in maize fields. When a plant produces its own protection against pests, developers are required to obtain an Experimental Use Permit (EUP) from the EPA, which specifies the conditions of the experiment. It should be noted that assessing plants that produce their own pest protection leaves the EPA with a limited role in overseeing the environmental impact of GMOs. The agency has complained regularly about this limitation since 1986, but with only limited success (Nap et al. 2003: 9–10; Bernauer and Meins 2003: 664; MacKenzie 2000: 43–45).

When it comes to the commercialization of GM foods, the American policy appears even more permissive. The Food and Drug Administration (FDA) has the premarket responsibility to ask for an assessment of the health effect of any substance added to food. The agency also specifies labeling requirements for food products. Finally, the FDA possesses the postmarket authority to have any product suspected of posing health risks removed from supermarkets. However, the FDA's postmarket and labeling authority has not been exercised specifically for GM food, the agency's intervention being confined to the premarket stage (Bernauer and Meins 2003: 663–64). However, the involvement of the FDA was unobtrusive even at the premarket stage. After the review of a single GM tomato, the Flavr Savr®, the FDA concluded that GM food was substantially equivalent to conventional food. After this review in 1992, the agency released its Policy on Foods Derived from New Plant Varieties, which relaxed premarket regulatory requirements for developers, establishing a voluntary notification procedure for GM food prior to commercialization. This policy counts on the legal responsibility of developers to market safe products to relieve the FDA from conducting comprehensive safety assessments for each GM food. The FDA's approach to its premarket responsibilities regarding GM food exemplifies, once again, the permissiveness of the American policy. In fact, the notification procedures for GM cultivars and food in the United States make this country's GMO policy even more permissive than that of Canada, the only country covered in this book that also has a permissive GMO policy.

## EXPLAINING THE AMERICAN ART AND GMO POLICY DESIGNS

Thus far, we have shown that both the ART and GMO policy designs are permissive. In the following sections, we argue that these permissive policies were the result of two very distinctive policy-making processes. First, the permissive ART policy stems largely from nondecisions in the legislative and executive branches, caused by major disagreements on an appropriate policy among key policy-making actors. Second, and in contrast to the ART policy design, the GMO policy results from explicit decisions to intentionally design a permissive policy. Such decisions were possible because key GMO policy actors shared the view that biotechnology is safe and carries an enormous economic potential. The two different policy-making processes leading to permissive policies in the ART and GMO sectors are presented below through an examination of policy networks and institutional arrangements. Internationalization has had so little influence on ART and GMO policy decisions in the United States that we decided that it did not deserve a separate subsection, as it did in the other chapters of this book.

## Policy Networks

Policy networks are defined by the nature of the relationships actors sustain when they participate in policy-making. Networks normally involve civil society as well as state actors. Two characteristics of policy networks influence policy designs. The first characteristic is the extent to which networks actors have cohesive policy beliefs and ideas. The more cohesive that a network is around a particular policy idea, the more predictable are the policy outcomes. British scholars (Rhodes and Marsh 1992) have chosen the term "policy communities" to describe small and cohesive policy networks, whereas large and diverse networks are known as "issue networks." As indicated in the first chapter of this book, network cohesion often proves insufficient to translate ideas into policies. More often than not, ideas are translated into policies when networks also provide a tight interconnection with or access to decisive state actors (Bressers and O'Toole 1998). A policy community that includes an administrative agency that possesses key policy responsibilities, for instance, is more likely to obtain the translation of its ideas into policy than one that does not include a similarly decisive actor. As hypothesized in the first chapter, we expect permissive biotechnology policies to stem from a policy community of target groups who naturally hold pro-technology beliefs and include key administrative or state actors. The presence of a network with these characteristics clearly serves to explain US GMO policy. As discussed below, however, issue networks also shape policy-making in ways that have encouraged permissive policies in the ART sector, a conclusion that goes against our hypothesis from the first chapter regarding issue networks. Pro-technology actors can also participate in issue networks, harming the efforts of biotechnology opponents who normally use these networks.

The USDA, Vogel (2002) explains, had a prominent role in the formulation of the Coordinated Framework for GMOs. The EPA argued in favor of a process-based regulatory approach whereby the agency would have large responsibilities in assessing the environmental impact of anything produced through rDNA technologies. The EPA's reasoning was that rDNA technologies were sufficiently distinctive from conventional biotechnologies that they might pose distinctive risks and therefore require specific expertise for their assessment and management. This reasoning has in fact informed much of the regulatory arrangements of European countries. As we saw above, the OSTP decided in 1986 in favor of the USDA's product-based approach.

In the mid-1980s, very few civil society actors were interested in GMO policy; only the research community and the emerging biotechnology industry displayed any interest. It is therefore unsurprising that the Coordinated Framework was very much the product of bureaucratic politics (i.e., a policy-making process confined within the executive branch and where interested departments competed to obtain the endorsement of their preferred way to address new problems from decision-makers). Bureaucratic politics is at the basis of the hypothesis presented in the first chapter regarding the policy influence of administrative organizations. As Allison and Zelikow (1999) argue, the importance of bureaus and their resources have a decisive influence on the outcome of bureaucratic politics. In the mid-1980s, the EPA was still a recently created agency. In contrast, the USDA had a long history and an enviable reputation (Carpenter 2001: chapters 6–7; Montpetit 2003: 86–87). Within the executive branch, the Secretary of Agriculture carried much more weight than the EPA's administrator. Moreover, the resources of the USDA, including its wide network of offices across the United States, left little doubt that the agency had sufficient capacity to effectively exercise new responsibilities. In comparison, the EPA's capacity appeared limited (Sheingate 2006). It is therefore unsurprising that the USDA prevailed in the bureaucratic politics that marked the development of the Coordinated Framework in the 1980s.

Even more surprising is that the Coordinated Framework remained unchanged since 1986 and that subsequent decisions, notably about voluntary notification, all went in the direction of even greater permissiveness. GMO policy has since 1986 emerged as a topic of debate within civil society; the EPA has increased its jurisdiction and influence within the executive branch; and it has become increasingly apparent that the USDA has had conflicting missions regarding GMOs. As the National Academy of Science (2002: 19) stated in a recent report, "there have been concerns that an agency [USDA] with the mandate to promote U.S. agriculture may not be able to objectively assess the safety of new products of agricultural biotechnology." If the policy remained so stable in this changing environment, we argue, it is because a policy community, that is a close and cohesive network of actors, formed around the USDA and shared the USDA's perspective on GMOs. This policy community locked in, effectively, a permissive policy trajectory (Pierson 2004).

Policy networks are informal policy-making arrangements and it is therefore difficult to distinguish clearly who is and who is not a member. As an indication of who is included in the GMO policy community, we examined all the public hearings held since 1997 by the House and Senate committees that have a potential interest on GMOs, whether or not they were dealing with specific GMO bills. The House's committees are Agriculture; Energy and Commerce; Resources; and Science. The Senate's committees are Agriculture, Nutrition and Forestry; Commerce, Science and Transportation; Energy and Natural Resources; and Environment and Public Works. We identified seven hearings related to GMOs, of which four were held by the House (3) and the Senate's (1) agriculture committees. The other three hearings were held by the House's science committee (basic research subcommittee). Interestingly enough, Congress's environmental committees have not held hearings on GMOs since 1997. The hearings were clearly dominated by industry groups, such as the American Farm Bureau Federation, the Biotechnology Industry Organization, the National Cotton Council, the American Soybean Association, the American Crop Protection Association, the Council on Agriculture Science and Technology, National Corn Growers Association, the American Seed Trade Association, the Institute of Food Technology of America, the Grocery Manufacturers of America, the National Food Processors Association, the National Agriculture Biotechnology Council, and the National Science Foundation.

Only one consumer group (Consumers Union, only preoccupied by labeling) and one environmental group (the Environmental Defense Fund) participated in the hearings of the science committee. It is therefore fair to conclude that groups such as Greenpeace, Friends of the Earth, the Organic Consumers Association, Public Citizen, among others, who publicly expressed concerns about biotechnology, albeit not in a particularly cohesive manner, are at best loosely interconnected with decisive state actors. This latter group of actors forms an issue network whose activities have essentially targeted civil society. The characteristics of the Canadian network of environmentalists and consumer groups are similar, as chapter 4 reveals. Targeting civil society is a perfectly justifiable strategy for groups opposing

GMOs, as Americans remain largely unconcerned about them. Bernauer and Meins (2003) speak of a low level of "public outrage," which is defined by the combination of low public concerns and high trust in regulatory authority. In contrast to European countries, public opinion did not exert pressure on American policy-makers to adopt stringent policy measures regarding GMOs. Americans have not radicalized their perception of GMOs (Bernauer and Meins 2003). Therefore, groups opposing GMOs focus their resources towards encouraging public outrage rather than directly influencing policy-makers. This strategy has had little success thus far in the United States, leaving the industry network undisturbed in its domination over the American GMO policy design.

The discourse of the actors of the industry network is highly cohesive in comparison to environmental and consumer groups. Indeed, the actors of the industry network have all argued in favor of the Coordinated Framework, suggesting it constitutes the best approach to regulating GMOs. To bolster the legitimacy of the Coordinated Framework, these groups frequently claim that the American approach is "science-based." These actors have not hesitated to argue that stringent approaches, such as those of European countries, involve motivations that cannot be justified in scientific terms. And in contrast to the UK, where representatives of food distributors became cautious about permissive policies (see chapter 5), all the actors of the American industry network respected this line of argument. In a hearing of the House Committee on Agriculture, held in the spring of 1999, a representative of the American Farm Bureau Federation made a statement that clearly illustrates the policy ideas and beliefs held by the actors of the industry network:

> Back in the 1980s, a system was reviewed in detail by the country's top scientists and the public before it was formally approved. The question in the 1980s concerned the need for a new super agency to monitor biotechnology use, especially for medicines and the answer was no. Existing agencies were given new and formal tasks, including the USDA, EPA, and the FDA. My conclusion is that no other country requires such comprehensive tests. The proof is in the absence of problems today. The current system is fully adequate, now and for the future. It is supporting biotechnology advances. . . . Circumstances have appeared to suggest either a new regulatory approach or new regulatory bodies are needed. The wisdom of the 1986 rejection of a super agency to regulate biotechnology has been reaffirmed through the 1990s. The evidence is compelling. Not only are there effective rules and watchdog agencies in the United States, but governments around the world evaluate and test new biotechnology products.[4]

Hearing this argument over and over again from the members of the industry networks, Congress unsurprisingly decided against adopting a GMO act or amending existing acts to increase the restrictiveness of the current policy. In a report prepared by the Subcommittee on Basic Research of the House's Committee on Science, published in 2000, called *Seeds of Opportunity* and presented as a "summation of the findings of a series of three hearings" held in 1999, the American policy regarding GMOs is presented as presenting no particular problem. Indeed, the subcommittee made the following observations in its report:

- "The concept of 'substantial equivalence' in the regulation of foods developed using agriculture biotechnology is scientifically sound and provides a useful historical baseline for judging safety";
- "There is no scientific justification for labeling foods based on the method by which they are produced. Labeling of agricultural biotechnology products would confuse, not inform, consumers and send a misleading message on safety";
- "Federal regulations should focus on the characteristics of the plant, its intended use, and the environment into which it will be introduced, not the method used to produce it";
- "Much of the opposition to agricultural biotechnology is politically motivated, not scientifically based" (Subcommittee of Basic Research 2000: 52–65).

Even the Environmental Protection Agency (EPA), the only agency that had been at some point critical about the product-based approach (Vogel 2002; Jasanoff 1995; Bernauer and Meins 2003), now seems to adhere to the science-based argument with few reservations. In a presentation before the House Committee on Agriculture in 2003, the EPA's administrator stated that: "EPA believes that the regulatory system is based on the most rigorous scientific information available, is credible, is defensible, and will serve to protect the environment and public health, and can evolve to meet the important challenges that lie ahead."[5] The industry network has locked in, powerfully, the permissive GMO policy trajectory adopted in the 1980s.

While the GMO sector is characterized by the presence of a cohesive policy community that provides industry actors with a tight interconnection to decisive state actors, issue networks dominate the ART sector (Hula 2005). One of the major differences between GMO and ART networks is the absence of a unique position among actors who value scientific advances. In the GMO sector, industry, and indeed the larger part of the research community, endorse the regulatory approach put in place through bureaucratic politics in the 1980s. In contrast, the ART research community is divided on fundamental policy issues, thereby fragmenting the pro-technology network. If almost all actors agree that the cloning of a human being for reproductive purposes is unreasonable, severe disagreements about the ethics and the necessity to clone for stem cells, so-called therapeutic cloning, divide the network. While some scientists urge Congress to pass a law prohibiting all forms of cloning, others argue that such a decision would significantly harm research. For the former group of scientists, cloning for stem cells will contribute to the advancement of knowledge of cloning in general and thus encourage the use of cloning for other purposes, including reproduction. Several of these scientists argue that stem cell research can be carried out effectively on adult stem cells and therefore embryonic research for that purpose is unnecessary. Other scientists vehemently oppose this view, arguing that embryonic stem cells can develop into a wider range of tissues than adult stem cells. In addition, they believe that cloning for stem cells increases the success rate of treatments significantly, as it allows the cells to come from the organism of the patient.

These divergent positions among the scientific community are reflected in the congressional hearings. Groups like the Biotechnology Industry Organization (BIO) and the American Society for Reproductive Medicine (ASRM), two of the largest groups representing both the industry and the research community, have pleaded for a clear distinction between therapeutic cloning and reproductive cloning to prevent harming stem cell research. However, while the former group agrees that reproductive cloning should be banned, the latter argues that federal legislation is unnecessary and that the research community is fully capable of regulating itself.

Pro-life groups are certainly united in the belief that all forms of embryo research should be prohibited. Groups such as the National Conference of Catholic Bishops, the United Methodist General Board of Church and Society, and the Culture of Life Foundation all testified before Congress that life begins at conception and therefore destroying embryos is no different than committing murder. Even actors who do not contribute to the moralization of American politics through religion have questioned the ethics of embryonic research. Some of them have called for a complete ban of all forms of cloning. For example, Leon Kass, chairman of the President's Council on Bioethics, and Francis Fukuyama, a well-known social scientist, believe it would be difficult to prevent reproductive cloning if the therapeutic form is permitted. Making the picture even more complex, some minor religious groups have taken the exact opposite stance. The Raelien Cult, and its industry arm Clonaid, in a notorious testimony before Congress, pleaded for total freedom to conduct research on cloning. As Hula (2005) argues, opponents to ART often make strange bedfellows. Groups opposed or concerned about biotechnology in the ART sector surely are sharply divided.

The extent of the divergence in views, and not only among opponents, creates a difficult context for decision-making. Further complicating the process is the fact that actors from all sides in this debate are effectively connected with decisive decision-makers. In fact, researchers, industry and religious organizations all have their points of entry to Congress as well as in the executive branch. This is in stark contrast with the loose interconnection that is detrimental to environmentalists and consumers in the GMO sector. It is unsurprising then that none of the 45 bills relating to ART, introduced in Congress between 1996 and 2004, was adopted and that presidents have had great difficulty in opening up the public purse for embryo research.

This discussion of policy networks, summarized in table 3.3, confirms that a policy community of pro-biotechnology target groups, which have a tight interconnection with decisive administrative and state actors, encourages the formulation of permissive policy designs. These conditions were indeed present in the GMO sector. The industry network faced an issue network, lacking cohesion and that provided opponents to GMOs only limited influence on decisive policy-makers. In light of this, we correctly anticipated that the GMO policy design is indeed permissive. This discussion of policy networks also confirms that the presence of a policy community of target groups is not the only route to a permissive policy. In fact, issue networks can encourage nondecision. This occurs when the networks provide actors with different policy perspectives and a tight interconnection with or access to a variety of decisive state actors. When new technologies appear, the incapacity to make decisions encourages a laissez-faire or permissive

**Table 3.3.   Contrasting ART and GMO Network Conditions**

|  | Pro-technology networks | | Networks of actors concerned about technology | |
|  |  |  |  |  |
|  | Medical network (ART) | Industry network (GMO) | Environmental/ Consumer network (GMO) | Network of Nonmedical opponents (ART) |
|---|---|---|---|---|
| Cohesion | Low | High | Low | Low |
| Interconnectedness | Tight | Tight | Loose | Tight |

policy orientation. This is precisely what occurred in the ART sector in the United States. To better understand this nondecision policy-making process, it is useful to examine decisive state actors and how the system of fragmented governance prevailing in the American institutional context makes it difficult to make decisions.

## Country Patterns

Two aspects of the American polity shed light on the ART and GMO policy-making processes: party politics and the fragmentation of the American political system. In the first chapter, we hypothesized that fragmented governance, when the possibility of coordination among fragmented policy-making arenas is absent, can encourage nondecision. In the GMO sector, we argue that the degree of agreement between the Republican and the Democratic parties on the appropriateness of biotechnology cancels out the typical effects of fragmentation to encourage decision-making. Party politics work in the same direction as policy networks in favor of permissive GMO policy decisions. However, party politics, just like issue networks, reinforces the system's fragmentation to encourage nondecision in the ART sector. This is consistent with the hypothesis formulated in chapter 1 regarding fragmented governance.

Like British politics, two political parties dominate American politics: the Republican Party and the Democratic Party. Unlike British politics, however, the two American parties are known to have similar positions on several policy issues. In addition, the elected members of the two American parties are not compelled to adhere to the party line during congressional votes. In theory, these characteristics of American party politics reduce the significance of the label of the governing party as the predictor of policy decisions, although Klingemann, Hofferbert and Budge (1994: chapter 8) suggest otherwise. We argue that party labels are poor predictors of policy designs in the United States, a situation that, as other chapters have shown, also prevails in several places. This is not to say, however, that party politics does not matter.

The issue of GMOs does not sharply divide Democrats and Republicans, as both parties have been supportive of the Coordinated Framework. If one might expect more sensitivity for the environment and consumers among Democrats than among Republicans, there are simply very few members of Congress who demand stricter assessments of GMOs, and they do not have sufficient support to obtain decisions that would make the American GMO policy design more restrictive. For example, Democrats had an opportunity to strengthen GMO policy between 1993 and 1995, as they controlled the executive and legislative branches. However, only one bill (H.R. 2169) related to GMOs, one which offered moderate reforms on labeling, was introduced during the 103th Congress of 1993–1995 by Democrat Gerald Kleczka. The bill never stood a chance of becoming a law, gathering insufficient support among Democrats. Far from considering more restrictions, the Democratic administration continued the streamlining of the regulatory process during its eight years in the White House (1993–2001), with little opposition from Washington policy-makers. Even Al Gore Jr., who was given environmental responsibilities by Clinton and who had advocated tougher GMO regulations in the past (Sheingate 2004: 7), did not move forward on this matter.

The similarity of views held by the two parties on GMOs has significant policy effects. As we explain in greater detail below, the fragmentation of the American political system will often create a difficult decision-making environment, to a point where nondecisions are frequent. In the GMO area, decision-making was not difficult. Not only is this sector characterized by a cohesive policy community, both parties share the view that the Coordinated Framework, adopted in the 1980s, provides an adequate regulatory environment. In other words, Congress, which often opposes the policies of the executive branch, has been supportive of the administration's GMO policy. The conclusions of *Seeds of Opportunity*, the congressional report cited above, leaves little doubt about the extent of the support that the executive branch receives from Congress.

Divisions between parties over ART may initially appear to be stronger. Embryo research was in fact an issue in the 2004 presidential campaign and both candidates expressed conflicting views. While George W. Bush, the Republican candidate, made it clear that he disliked embryo research, John Kerry, the Democratic candidate, argued that the federal government should create better conditions for embryo research because it can help patients suffering from terrible diseases. In addition, it is often said that the Republicans owe their electoral successes to the Christian Right Coalition, which opposes embryo research on moral grounds. The reality, however, is not this simple. It should be recalled that it was Bush who allowed the use of public funds to carry stem cell research during his first term, although under strict conditions, after Clinton had maintained the ban during his two terms in office. In 2005, after a clear victory of the Republican Party in the 2004 congressional elections, two ART bills were debated and gathered significant support in Congress. Surprisingly enough for a Republican-dominated Congress, the two bills do not prohibit stem cell research, but encourage it. One is an uncontroversial bill (H.R. 2520) to permit the extraction of stem cells from umbilical cords. More significant is the Stem Cell Research Enhancement Act (H.R. 810), which loosens the conditions under which the NIH can fund stem cell research under the conditions put forward by Bush in 2001. This second bill was adopted by the House in the spring of 2005 and, in February 2006, was still awaiting floor action in the Republican Senate. What is delaying the bill in the Senate is not so much the lack of support for it among senators, as it is Bush's announcement that he would veto it. All this is to say that party labels do not inform ART policy more than they did inform GMO policy designs. However, this discussion also suggests that internal divisions on ART policy plague both political parties, while they were relatively united in their support for a permissive GMO policy.

The House vote on the Stem Cell Research Enhancement Act of 2005 illustrates the internal party divisions on ART, and in particular the division within the Republican Party. Indeed, 50 Republicans voted with 187 Democrats and 1 independent member in support of the bill. Meanwhile, 14 Democrats voted with 180 Republicans against the bill. This historic vote was preceded in 2004 by a letter signed by 206 House Representatives, including 36 Republicans, urging Bush to revise the 2001 conditions for the NIH; 58 senators, including 14 Republicans, signed a similar letter.

Discussions within the executive branch prior to the decision, announced by Bush in 2001, to partially lift the ban on the use of public funds to carry stem cell research also illustrate the internal party division and the difficulties it created for decision-making. In fact, this decision was not simply the translation of a party line into policy; it was the result of a compromise among actors who had diverging views on this issue. As Tommy G. Thompson, State Secretary for Health, told the *Washington Post* prior to the policy announcement: "Hopefully we'll come up with a decision that's going to allow for the continuation of research, which is very important, and at the same time take into consideration the legal and the ethical questions that have to be considered" (Connely and Weiss 2001). In other words, some Republicans within the executive branch were against the destruction of embryos and some supported embryonic research.

Although a compromise within the executive branch was possible in 2001, party division usually encourages nondecisions. As alluded to above, the difficulty of making decisions is often presented as an inherent characteristic of the American political system. The separation of powers between the executive, the legislative and the judiciary branches, as well as the division of powers between the federal and state governments, creates multiple sites of decisive policy-making. For Tsebelis (1995), these sites provide effective veto points to veto players. Everything else being equal, Tsebelis (1995) would argue, it is more difficult for anyone to stop in its tracks a policy supported by a prime minister in the Westminster system of the UK than it is to block a presidential policy, thanks to Congress and the courts in the United States. Adding to this logic of veto points is the divisions within political parties. If the Republican Party was not currently divided on ART, decision-making conditions in this area would be favorable in the short term, since the Republican Party controls the presidency and Congress. Internal party division increases the likelihood that both pro-biotechnology actors and actors worried about biotechnology develop tight interconnections with decisive state institutions. Combined with the fragmentation of the American political system, party division ensures that actors worried about technology can veto initiatives aimed at increasing the level of public funding for embryonic research as much as it ensures that pro-biotechnology actors will veto any policy to prohibit embryonic research.

In short, the level of agreement between and within the Republican and Democratic parties on a permissive GMO policy prevented the development of tight interconnections between actors worried about biotechnology and decisive state actors. In contrast, internal party divisions provided interconnections with state institutions to actors on all sides of the ART policy debate. Unsurprisingly then, Congress was unable to agree on the forty-five ART bills it received before 2004. If one bill currently stands a chance of obtaining the approval of both houses of Congress, the Republican president announced he would veto his Republican Congress, in what is generally treated as a context of undivided government. When party divisions combine with the fragmented institutions of the United States, decisions are difficult to make.

## CONCLUSION

Permissive policy in the GMO and ART sectors may come as no surprise to some observers of American biotechnology, who view these policies as being consistent with the faith that policy-makers put in science. In contrast, others may find it surprising that this country has biotechnology policies that allow scientists "to play god," given the current religious context of the United States. This chapter suggests that the consistency between the ART and GMO policies is a matter of coincidence, both policies having been governed by different processes and different actors.

The forms of policy networks, which contribute to explaining the two permissive policy designs, sharply diverged. A policy community of potential target groups, naturally favorable to a permissive policy, dominated the GMO sector. In a context where both parties endorsed the permissive policy decided in the 1980s, it was easy for the actors of this community to develop interconnections with decisive state institutions, thereby locking in the permissive policy trajectory. In contrast, the ART sector is characterized by sharp divisions within issue networks, divisions matched by those within political parties, especially the Republican Party. Under such divisive conditions, policy initiatives often stumble upon veto players. Because the American institutional context provides several veto points, nondecisions have characterized the ART sector. Therefore, permissiveness in the ART sector does not stem from deliberate decisions, such as in the GMO sector, but from the absence of decision.

The conclusions of this chapter are only partially consistent with the hypotheses presented in the first chapter. The presence of a community of target groups, also involving administrative agencies whose preferences match those of the target groups, had the expected effect on GMO policy. However, the issue network did not enable opponents or actors with concerns about biotechnology to press effectively for the adoption of a restrictive policy. The network, nevertheless, provided a diversity of actors with interconnection and access to decision-makers, a situation made possible by the American system of fragmented governance. If one adds party division to this picture, the exercise of veto over any policy change becomes unsurprising. In absence of a policy community and of cohesive political parties, fragmented governance and the veto it enables play a key role in the explanation of the ART policy design.

It should be recalled that internationalization, whether through policy transfers or international rules, played a negligible role in the explanation of both ART and GMO policies in the United States. More often than not, American policy-makers ignore international rules when they make their own domestic policies (Montpetit 2005). This being said, American policy-makers devote efforts to influencing the development of international rules and do not hesitate to use international institutions to influence the policies of other countries, as illustrated by the US challenge of the European Union's GMO policy. Again, such activities have very little bearing for the development of American biotechnology policies.

## NOTES

1. http://stemcells.nih.gov/policy/taskForce/, consulted on April 5, 2005.

2. Testimony before Senate by the director of NIH, www.senate.gov/~appropriations/hearmarkups/record.cfm?id=204165.

3. Section 263a-2—Certification of embryo laboratories.

4. Testimony before Agricultural Committee of the House of Representatives on March 3, 1999; http://commdocs.house.gov/committees/ag/hag1066.000/hag 1066_0f.htm.

5. Testimony before the House's Committee on Agriculture, June 17, 2003.

## REFERENCES

Allison, G., and P. Zelikow. 1999. *The Essence of Decision: Explaining the Cuban Missile Crisis, Second Edition.* New York: Longman.

Bernauer, T., and E. Meins. 2003. Technological Revolution Meets Policy and the Market: Explaining Cross-National Differences in Agricultural Biotechnology Regulation. *European Journal of Political Research* 42 (5): 643–83.

Bressers, H. Th. A., and L. J. O'Toole Jr. 1998. The Selection of Policy Instruments: a Network-based Perspective. *Journal of Public Policy* 18 (3): 213–39.

Carpenter, D. P. 2001. *The Forging of Bureaucratic Autonomy: Reputations, Networks, and Policy Innovation in Executive Agencies, 1862–1928.* Princeton: Princeton University Press.

Cibelli, J. B., R. P. Lanza, and M. D. West, with C. Ezzell. 2002. The First Human Cloned Embryo. *Scientific American* 286 (1): 42–9.

Connely, C., and R. Weiss. 2001. Stem Cell Research Divide Administration. Thompson Expresses Optimism That a Compromise Will Be Reached Soon. *Washington Post,* June 12, 2001.

Goggin, M. L., and D. Orth. 2004. The United States: national talk and state action. In *Comparative Biomedical Policy. Governing Assisted Reproductive Technology.* Ed. I. Bleiklie, M. L. Goggin, and C. Rothmayer. Pp. 82–101. London: Routledge.

Gunning, J., and V. English. 1993. *Human in Vitro Fertilization: A Case Study in the Regulation of Medical Innovation.* Dartmouth: Ashgate.

Hula, K. W. 2005. Dolly Goes to Washington: Coalitions, Cloning and Trust. In *The Interest Group Connection: Electioneering, Lobbying and Policymaking in Washington.* Ed. P. S. Herrnson, R. G. Shaiko, and C. Wilcox. Pp. 229–48. Washington: CQ Press.

Jasanoff, S. 1995. Product, Process, or Program: Three Cultures and the Regulation of Biotechnology. In *Resistance to New Technology.* Pp. 311–31. Ed. M. Bauer. Cambridge: Cambridge University Press.

Klingemann, H.-D., R. I. Hofferbert, and I. Budge. 1994. *Parties, Policies and Democracy.* Boulder: Westview Press.

MacKenzie, D. J. 2000. *International Comparison of Regulatory Frameworks for Food Products of Biotechnology.* Paper prepared for The Canadian Biotechnology Advisory Committee Project Steering Committee on the Regulation of Genetically Modified Foods, Ottawa.

Montpetit, É. 2005. A Policy Network Explanation of Biotechnology Policy Differences between the United States and Canada. *Journal of Public Policy* 25 (3): 339–66.

Montpetit, É. 2003. *Misplaced Distrust: Policy Networks and the Environment in France, the United States and Canada*. Vancouver: UBC Press.

Nap, J. P., P. L. J. Metz, M. Escaler, and A. J. Conner. 2003. The Release of Genetically Modified Crops into the Environment. *The Plant Journal* 33 (1): 1–18.

National Research Council. 2002. *Environmental Effects of Transgenic Plants: The Scope and Adequacy of Regulation*. Washington: National Academy Press.

The President's Council on Bioethics. 2004. *Monitoring Stem Cell Research*. Washington, DC: Government Printing Office.

Rhodes, R. A. W., and D. Marsh. 1992. New Directions in the Study of Policy Networks. *European Journal of Political Research* 21: 181–205.

Sheingate, A. D. 2006. Promotion versus Precaution: The Evolution of Biotechnology Policy in the United States. *British Journal of Political Science* 36: 243–68.

Stone, D. 1997. *Policy Paradox: The Art of Political Decision Making*. New York: W. W. Norton & Company.

Tsebelis, G. 1995. Decision Making in Political Systems: Veto Players in Presidentialism, Multicameralism and Multipartyism. *British Journal of Political Science* 25 (3): 289–325.

U.S. Congress. House Committee on Science. Subcommittee on Basic Research. 2000. *Seeds of Opportunity: An Assessment of the Benefits, Safety, and Oversight of Plant Genomics and Agricultural Biotechnology*, 106th Cong., 2nd sess.

U.S. Department of Health and Human Services. 2004. *2002 Assisted Reproductive Technology Success Rates*. Atlanta, Georgia: Center for Disease Control and Prevention,

Vogel, D. 2002. Ships Passing in the Night: GMOs and the Politics of Risk Regulation in Europe and the United States. Working paper of The Center for the Management of Environmental Resources.

# THE CANADIAN KNOWLEDGE ECONOMY IN THE SHADOW OF THE UNITED KINGDOM AND THE UNITED STATES

*Éric Montpetit (Université de Montréal)*

Canada is, population wise, a small country that is often viewed as trying to play in the big leagues. In terms of the knowledge economy, the government of Canada (1998: 10–11) insists that the country ranks third behind the United States and the United Kingdom in the global biotechnology market and that its policies work to "position Canada as a responsible, innovative world leader." However, playing in the big leagues, cynics might argue, will only benefit Canada's biotechnology industry and erode the country's regulatory standards in this sector to the level of competing countries with the lowest possible standards.

I argue in this chapter that the influence of the United Kingdom and of the United States on Canada's ART and GMO policy designs do not amount to an alignment with the policies of these two countries. While the United Kingdom did inspire some policy transfers in the area of ART Canada made far more restrictive ART policy choices. In addition, Canada adopted a slightly more restrictive policy in the area of GMO than the policies of its key trading partner, the United States, despite commercial pressure. In fact, the comparison of the Canadian ART and GMO policy designs reveals that policy networks, distinct in both areas, provide a convincing explanation. Party politics and federalism provide additional insights into Canadian biotechnology policy choices, while internationalization ends up playing, at best, a limited role. This argument is presented in detail after outlining the key distinctions between the Canadian ART and GMO policy designs.

## AN INTERMEDIATE ART POLICY

After ten years of policy development, the federal government adopted the Assisted Human Reproduction Act in March 2004. The act complies largely with the recommendations of the Royal Commission on New Reproductive Technologies, also known as the Baird Commission, published in 1993. Interestingly enough, the UK's Human Fertility and Embryology Act also complied with the recommendations of a public inquiry, which incidentally had also influenced the work of the Baird Commission. Unsurprisingly, several aspects of the British and Canadian acts are similar. Both the Canadian and UK acts are comprehensive in that they simultaneously deal with research and clinical applications; both contain prohibitions that are enforceable with criminal sanctions; and both provide for the creation of a regulatory authority responsible for the delivery of licenses to clinics, the approval of research projects resorting to human embryos, and for the collection of reproductive health-related data. However, some aspects of the Canadian act are more restrictive, hence its classification as an intermediate rather than a permissive ART policy design.

The key reason for classifying the Canadian ART policy design as an intermediate rather than a permissive design rests with the prohibitions related to embryo research (see table 4.1). In fact, the Assisted Human Reproduction Act prohibits the creation of embryos for research as well as therapeutic cloning. In contrast to British researchers, Canadian researchers are limited to using embryos leftover from the process of IVF treatments. Furthermore, human embryo research can be approved only to the extent that researchers are able to demonstrate that their research could not be carried out on animal embryos. Stem cell research is further constrained in that researchers must obtain the written consent of the suppliers of gametes. However, in contrast with the British Human Fertilisation and Embryology Act, the Canadian act does not specify the purposes for which embryo research can be conducted, leaving great discretion to the regulatory agency, the Assisted Human Reproduction Agency of Canada (AHRAC).

Indeed, the Canadian Assisted Human Reproduction Act creates an agency, the AHRAC, whose responsibilities will include the licensing, regulating and monitoring of clinics. Given that the agency is still in development, regulatory decisions on matters such as limiting prenatal diagnosis to a predetermined set of diseases or tissue typing have not yet been made. The act, however, makes clear that embryo screening for sex selection is prohibited as well as the commercialization of gametes. With regard to the commercialization of gametes, the Canadian policy is similar to that of the UK insofar as the AHRAC will be empowered to decide what constitutes reasonable expenses to be refunded to donors of gametes.

Most ART treatments in Canada are not covered by provincial health insurance schemes. However, the Canadian Assisted Human Reproduction Act states, in its introductory principles, that "persons who seek to undergo assisted reproduction procedures must not be discriminated against, including on the basis of their sexual orientation or marital status." Patients are thus protected against potential clinical discrimination based on their sexual orientation and marital status, which means that access to fertility treatments is only constrained by the financial capacity of patients. Among the countries covered in this book, none have a policy with such a progressive clause that protects access to categories of citizens.

In summary, Canada adopted in March 2004 a policy that is rather permissive from the viewpoint of patients. For fertility clinics, the act specifies some restrictive conditions, notably with regard to gamete donation. Most conditions related to fertility treatments, however, will be contained in regulations that are still in development. The conditions that make the Canadian policy design more restrictive than that of the United Kingdom are the prohibitions related to embryo research. In contrast to British researchers, Canadian researchers cannot create embryos for research or clone embryos to obtain stem cells. Embryo research, including stem cell research, can only be carried out on embryos created for procreation, but which were never implanted into the womb of a woman; this represents a significant change from the situation that had prevailed in Canada prior to 2004 (Montpetit et al. 2004). Table 4.1 summarizes the intermediate policy design of Canada.

## PERMISSIVE GMO POLICY

Canada, like the United States and in contrast to European countries, chose in the 1980s a product-based policy toward GMOs (Jasanoff 1995). Instead of having laws and specialized agencies to regulate GMOs, Canada resorts to the existing laws and agencies that have had the responsibilities for the approval of conventional seeds, pesticides or food. Canada does not have an advisory body, such as Britain's ACRE, whose responsibility is to provide the government with guidance in the approval of GM cultivars. Instead, the Canadian Food Inspection Agency (CFIA) approves or rejects GM cultivars along with conventional cultivars, using similar practices and regulations. In

Table 4.1.   Current Canadian ART Policy Design

|  | Research | Practice of fertility treatments | Access for patients |
|---|---|---|---|
| Policy Design | - Research on leftover embryos permitted<br>- Creation of embryos for research and therapeutic cloning prohibited | - Commercialization of gametes and surrogacy prohibited<br>- Expenses related to gamete donation can be refunded, given that the donor provides a receipt<br>- Loss of income is considered an acceptable expense for surrogacy<br>- Gamete donations have to be anonymous<br>- Sex selection prohibited, but other prenatal diagnosis unregulated for the time being | - Access unlimited<br>- The act provides against clinical discrimination against unmarried and same sex couples<br>- No public insurance coverage for most treatments |
| Type | Intermediate | Intermediate | Permissives |

addition to representing a significant contrast with European countries, this product-based approach is also different from the approach that has prevailed in the area of ART in Canada. Again, instead of relying on existing laws and regulatory mechanisms pertaining to medical practices and research, the Canadian government adopted a law and created an agency specifically to regulate ART research and clinical applications. An additional distinction between the Canadian GMO policy and its counterpart in the ART area rests in the permissiveness of the former. Policy permissiveness characterizes research in the GMO area, while policy restrictions in the ART sector apply precisely to research.

The Canadian GMO policy design was formalized in 1993 in the *Regulatory Framework for Biotechnology* (Yarrow 2001). The framework required amendments to the Feeds Act, the Fertilizers Act, the Seeds Act, the Health of Animals Act and the Plant Protection Act as well as the related regulations. These legislative and regulatory changes specified that novel plants and food would require risk assessments that would be governed by the Canadian Environmental Protection Act (CEPA) and/or the Food and Drugs Act (FDA). The CEPA provides the foundation of an environmental risk management process and the FDA offers the foundation of a health risk management process. The application of CEPA with regard to novel cultivars, including GMOs, is not the responsibility of Environment

Canada, but rather of the Canadian Food Inspection Agency and to a lesser extent the Pest Management Regulatory Agency (PMRA). Health risks of any GMOs likely to enter the food chain, including GM cultivars, are managed by Health Canada.

Therefore, biotechnology developers, who want to obtain an authorization to conduct field experiments in view of commercializing a GM cultivar, are required to notify the CFIA and comply with the Seeds Act, which was adjusted in 1996 to meet CEPA's standards. While the Seeds Act has traditionally provided protection against agronomic risks related to the introduction of new seeds into the environment, the adjustments of 1996 require developers to provide information on the toxicity of GM cultivars. In short, developers have to provide more information for novel cultivars than for conventional ones. However, developers can conduct field trials in Canada provided that they respect separation distances to prevent transfers of pollen to other plants as well as restrictions on the storage, use or disposals of tested crops. The CFIA only receives notification and carries out inspections to enforce these regulatory conditions. The PMRA can also be involved in a similar manner if the novel trait provides the plant with a direct protection against pest.

Field trials of GM cultivars or laboratory experiments on GM food products provide the information required by the CFIA and Health Canada to manage environmental and health risks. The CFIA and Health Canada's officials analyze all the information that the policy requires from developers in order to approve, approve conditionally or reject GM cultivars and foods on a case-by-case basis. By the fall of 2004, the CFIA had approved forty-nine GM cultivars and Health Canada fifty food products containing GMOs as suitable for commercialization in Canada. This process is more restrictive than the American process where most GM cultivars are fast-tracked and GM food considered safe since the 1992 health risk assessment of the Flavr Savr tomato (see chapter 3). However, Barrett and Abergel (2000) claim that more and more GM cultivars and foods escape regulatory approvals in Canada because regulators treat them as familiar or substantially equivalent to cultivars or foods already approved. The government of Canada (2001: 11), however, claims that it uses familiarity and substantial equivalence as comparative tools in thorough processes of risk assessment. The government suggests that it complies with the recommendation of the OECD and the World Health Organization never to use familiarity and substantial equivalence as decision thresholds.

Further illustrating the permissiveness of the Canadian GMO policy design is the absence of traceability as a potential regulatory issue. Moreover,

**Table 4.2.   Current Canadian GMO Policy Design**

| | Field trials | Commercialization and cultivation | Consumer and environmental protections |
|---|---|---|---|
| Policy Designs | - Experiments in fields generally allowed, but regulated | - GM cultivars and Novel foods approved for commercialization on a case-by-case basis | - Voluntary labeling for products containing GM traits representing more than 5 percent of total product composition |
| Type | Permissive | Intermediate | Permissive |

although the Canadian General Standards Board has regulated the labeling of products as genetically engineered since 2004, developers label on a voluntary basis only. And only those products whose composition contains GM ingredients representing 5 percent or more of the total weight can be labeled (Government of Canada 2004: 4–5). As a point of comparison, the European standard is 0.9 percent.

If one compares tables 4.1 and 4.2, the difference between the Canadian ART and GMO policy designs may not appear that striking. In comparison, the differences identified in chapter 5 between the British ART and GMO policy designs were much sharper. However, it is fair to suggest that the Canadian knowledge economy is also characterized by some policy contrasts. In the area of GMOs, policy-makers adopted a product-based approach, whereas in the area of ART, the approach is akin to a process-based approach. More importantly, the Canadian ART policy design constrains research while GMO research can be carried out relatively free of regulatory burdens. Conversely, licensed ART clinics can offer a wide range of fertility treatments, but GMO developers must obtain approval on a case-by-case basis in order to market their products. Nevertheless, I argue that the Canadian ART policy design is overall slightly more restrictive than the GMO policy design.

## EXPLAINING THE DIFFERENCE BETWEEN ART AND GMO POLICY DESIGNS

Policy networks offer a powerful explanation of the difference between the GMO and ART policy designs. Network situations in the two areas differ significantly and the differential impact of Canadian institutions add to the effects of networks to account for these differences. As explained at the end of this section, the internationalization of policy also had a differing degree of impact, but to a lesser extent.

## Policy Networks

Canada was inspired to intervene in the area of ART by the example of the UK. At the end of the 1980s, the Conservative government of Brian Mulroney set up a Royal Commission on new reproductive technology (RCNRT), which received a mandate largely similar to that of Britain's Warnock Committee. The RCNRT's report, published in 1993, was substantially analogous to the Warnock report, recommending some prohibitions but also setting up an agency to regulate ART. However, ART policy development in Canada took place between 1993 and 2004, a long period during which the issue lost much of the simplicity it had in 1980s Britain. The complexity of the issue is reflected in the network situation, which, as Table 4.3 suggests, does not reflect a simple opposition between medical and religious actors. Networks are more numerous in Canada than in the UK and the contrasts are not as stark in their cohesion as well as in the interconnectedness they provide to the state.

Demands for a public inquiry into new reproductive technology in Canada came from a coalition of women. In contrast, women's groups were largely absent from ART policy development in the UK and as a consequence, Mulkay (1997: chapter 6) argues, policy-makers there never saw ART as an issue of particular concern to women. In Canada, after the publication of the RCNRT's report, three networks were important for ART policy development. The women's network exerted more influence on the design of the Canadian Assisted Human Reproduction Act than the other two networks. Unsurprisingly then, the Canadian act contains restrictions

**Table 4.3.  Contrasting Network Situations between ART and GMO**

| | Pro-technology networks | | Networks of actors concerned about technology | | |
| | Medical network (ART) | Industry network (GMO) | Environmental/ Consumer network (GMO) | Religious network (ART) | Women's network (ART) |
|---|---|---|---|---|---|
| Cohesion | High | High | Low | High | High |
| Interconnectedness | Moderate | Moderate | Loose | Loose | Tight |

that are absent from that of the UK and acknowledges the importance of the issue to women. The Canadian Assisted Human Reproduction Act notably states that: "women more than men are directly and significantly affected by [ART] application and the health and well-being of women must be protected in the application of these technologies."

The influence of the women's network, however, was not straightforward and immediate. In a study of the Human Reproductive and Genetic Technologies Bill of 1996, the federal government's first attempt to legislate in this area, I have argued that the medical network contributed more than the other two networks to the failure of the bill in 1997 (Montpetit et al. 2004). What changed in Canada since 1997? As I will explain in the next section, a change in party politics appears to have facilitated the passing of the 2004 Canadian Assisted Human Reproduction Act. More importantly, however, the cohesion of the women's network and its interconnection with the state changed.

In 1997, the women's network was comprised of seven key actors (see Montpetit et al. 2004: 73) holding different views on what the federal government should do in the area of ART. For instance, the Feminist Alliance on New Reproductive and Genetic Technology insisted on the importance of regulations and expressed concerns about employing the criminal law as a means to prohibit certain practices. Criminal law, the group argued, could interfere with women's reproductive autonomy. Conversely, the National Action Committee on the Status of Women (NAC), then a key actor in ART policy development, was more hesitant to condemn the use of criminal sanctions. Adopting an "equality of results" perspective, NAC acknowledged that criminal sanctions could interfere with women's reproductive autonomy, but the group also insisted that too permissive a policy could fail to protect vulnerable women against abuses. With this 1997 position, NAC had already distanced itself significantly from the position it had adopted during the work of the RCNRT. Indeed, NAC was then a major opponent of ART, believing commercial interests were behind these technologies and that they would contribute to maintaining women's subordination to men (Montpetit, Scala, Fortier 2004: 145). Until 1997, the diversity of views within the women's network reduced its policy influence.

The network, however, changed drastically between 1997 and 2004. The Feminist Alliance on New Reproductive and Genetic Technology ceased to exist. NAC experienced severe difficulties during this period and stopped participating in ART policy development. In fact, when the House Standing Committee on Health began consultations for a draft bill in 2001, the women's network, with the exception of REAL Women of Canada (arguably closer to the religious network), was comprised of only service-oriented groups, often with a local base, and individual women previously involved with NAC and other groups. At first sight, this change may have appeared to imply a weakening of the network, but this was not the case.

The women's network in 2001 was in fact more cohesive and had suffered no loss in terms of its connection with decisive state actors. The service-oriented groups that formed the post-1997 network were primarily concerned with women's health. Positions on which practices to prohibit and which to regulate are thus solely informed by their medical impact on women and resulting children. As we saw above, prior to 1997, the network was primarily made up of advocacy-type groups that did not share a single focus. Moreover, several individuals who were involved in ART policy development within these advocacy groups prior to 1997 continued their involvement within the service-oriented groups or in an individual capacity after 1997. The personal contacts constructed since the early 1990s between individuals acting in women's organizations and policy-makers were thereby preserved. As an illustration, the Canadian Women's Health Network (CWHN) became a key actor within the women's network after 1997. Abby Lippman, who had previously been involved with NAC and the Feminist Alliance on New Reproductive and Genetic Technology, acted as a spokesperson for the CWHN during the ART hearings of the House Standing Committee on Health in 2001. Lippman's personal contacts therefore contributed to the tightness of the interconnection that the network had with state actors. In addition, the CWHN is tightly interconnected with Health Canada through the Women's Health Contribution Program, a government program administered by Health Canada. This program finances some of CWHN's health services. The connection to Health Canada, which is an enduring one, is significant because Health Canada had the responsibility for drafting the ART legislation. Cohesive and offering a tight interconnection, the women's network unsurprisingly influenced the design of the Canadian Assisted Human Reproduction Act.

In contrast, the composition of the medical network, formed by the prime targets of the restrictions contained in the act, remained stable, and actors forming the network did not change their views since 1997 (for an overview of the network, see Montpetit 2004: 73). Actors such as the Society of Obstetricians and Gynaecologists of Canada (SOGC) were always convinced that criminal sanctions were inappropriate, that medical self-regulation at the provincial level was adequate and that an eventual federal authority responsible for ART should have minimal regulatory responsibilities. This policy preference contrasts with that of the actors of the UK's medical network, who were unsatisfied with self-regulation in the 1980s and demanded the creation of a state licensing authority. In the next section, I will provide an explanation for this difference in policy preferences. For the moment, suffice it to say that support for this uncompromising position became increasingly difficult to obtain from state actors.

The interconnection between the medical network and the state prior to 1997 was through Parliament. The interconnection was diffuse and fragile since it rested on the relationship that members of Parliament (MPs) had developed with physicians in their constituency during the abortion debate of the early 1990s (Montpetit 2004: 75–76). This interconnection surely contributed to defeating the Human Reproductive and Genetic Technologies Bill in 1997. It should be underlined that the Human Reproductive and Genetic Technologies Bill, in contrast to the Canadian Assisted Human Reproduction Act, consisted only of thirteen criminal prohibitions. The key argument of the medical profession against the Human Reproductive and Genetic Technologies Bill was that the use of criminal law to regulate medical practices would tarnish the reputation of physicians, undermining the relationship of confidence they enjoy with their patients. MPs accepted this argument. Although the 2004 Canadian Assisted Human Reproduction Act contains criminal prohibitions, it is not strictly a criminal law act because it also provides for the creation of a federal regulatory agency with extensive responsibilities. Being opposed to the creation of a powerful federal agency, the medical profession argued, along with several provincial governments, that provincial self-regulatory devices were adequate. Federal MPs were not as inclined to accept a line of argument taking the provincial side in an intergovernmental competition, which is a constant preoccupation of policy-makers in Canada.

The health minister's decision to involve Parliament more directly into the preparation of the Canadian Assisted Human Reproduction Act helped to strengthen the relationship between MPs and Health Canada, the ministry that connected the women's network to the state. By the end of 2000, it was widely believed that Health Canada had not properly consulted Canadians on ART, prompting the health minister to present a draft bill to the House Standing Committee on health. This irregular procedure enabled MPs to make recommendations on the draft of a bill before it was introduced to Parliament. The procedure presented few risks to the bill itself, to the extent that the committee chair who had just been appointed was not close to opponents to the bill, notably the medical profession. It also provided MPs with a sense of greater involvement in an area in which they had felt that, since 1997, Health Canada had been making all of the decisions. The committee's recommendations did not in fact significantly alter the draft bill and helped to gain support for the work of Health Canada among MPs, who otherwise could have been on the side of the medical network.

In comparison to the UK, the religious network in Canada is highly cohesive in its vehement opposition to embryo research. The importance of this network, however, has never been very high and is in decline. The influence of this network in the mid-1990s rested with its interconnection to the state via the Reform Party. This interconnection was never very powerful as the Reform Party was a regionally based opposition party. With its base in Western Canada, where social conservatism is popular, and sixty seats in Parliament in 1997, the Reform Party was an effective channel for the ideas of the religious network on ART. In an effort to widen its regional base, however, in 2000 the Reform Party became the Canadian Alliance. In order to obtain votes in Eastern and Central Canada, the Canadian Alliance had to distance itself from social conservatism. Further distancing itself from its regional base in Western Canada, the Canadian Alliance decided in 2003 to merge with the Progressive Conservative Party. With a wider electoral base, the newly created Conservative Party of Canada is no longer as effective a channel for the ideas of the religious network, although the majority of its MPs are still from Western Canada.

The network situation prevailing in the area of GMO is far simpler. In table 4.3, I identify, just as in the UK, two networks pertaining to GMOs: an industry network and an environmental/consumer network. Differences between Canada and the UK, however, are stark. First, the environmental/consumer network forms at best an issue network, that is a network of actors who meet infrequently and who are consulted by the state only periodically. It is a challenge for any analyst to delimit precisely the boundary of such networks as actors change constantly. For example, Greenpeace, formerly an important actor in the network, decided to close its office in Ottawa in a strategic move to focus its energies on changing consumer behavior rather than directly influencing government. The government claimed to have consulted the following nongovernmental organizations on the issue of GMOs: the Consumers' Association of Canada, the Canadian Environmental Network, Pollution Probe, the Sierra Club, the Council of Canadians and the Canadian Environmental Law Association. This network consists of actors with very different points of view towards GMOs. The Canadian Environmental Network, for instance, takes no clear position but encourages participation in debates and consultations on environmental issues, including GMOs. Meanwhile, the Council of Canadians vehemently opposes GMOs and the Consumers' Association of Canada simply demands mandatory labeling. In any case, the concerns of these groups on the impact of GMOs find very little echo within the state. The Ministry of the Environment in Canada is a weak ministry (Doern and Conway 1994), which did not even retain the responsibilities over GMOs conferred by CEPA (Leiss and Tyshenko 2002: 325). Again, most environmental assessments of GMOs are conducted under the Seeds Act by the CFIA. This network provides very poor access to decisive state actors for opponents or actors concerned about biotechnology. In comparison to Britain's environmental network, the Canadian environmental/consumer network failed to exert any significant influence on the country's GMO policy design.

In contrast, the Canadian industry network, comprised of representatives of input suppliers, farmers, processors and distributors, is more cohesive than its British counterpart. Thus far, consumer campaigns in Canada, such as those conducted by Greenpeace, have had very little success. Unlike British food processors and distributors, therefore, the Canadian food industry does not view it as necessary to endorse more restrictive policy positions. The only exception is GM wheat. In 2001, Monsanto Canada began field experiments to obtain permission to commercialize a GM strand of herbicide-tolerant spring wheat. Groups such as the Council of Canadians began a campaign to sensitize farmers about possible negative reactions, if the GM wheat were approved, in countries, notably Japan, where consumers do not accept GMOs. This campaign prompted, in 2003, farm groups representing wheat producers, as well as the Canadian Wheat Board, to ask for more caution than usual in the risk assessment of this GM brand of wheat. As explained above, assessments of GMOs in Canada only relate to health and environmental risks; wheat farmers demanded also an analysis of the economic risks of accepting GM wheat. Wheat is the single most important crop in Canada, especially in Western Canada, and much of it is exported. Farmers feared that they would lose market share if GM wheat were planted and demanded that the government consider the economic impact before making a decision. Given the reaction of farmers, Monsanto decided to withdraw its GM wheat from experimental fields in May 2004. It should be underlined that this disagreement within the industry network only pertains to wheat. In the letter to the federal minister of agriculture demanding an economic risk assessment, the Canadian Wheat Board, along with the representatives of ten wheat-related organizations, clearly stated that they did not seek a general review of GMO regulations, but rather that their request pertained only to wheat. They added that they "support the concept of working collaboratively with the developer of a variety to deal with these issues."[1] In short, despite a disagreement over GM wheat, the cohesion of the industry network remains high.

The interconnection of the network to the state, initially tight, became looser in 1997 when the CFIA was created. As an illustration of the initial tightness, the basis of the current GMO regulatory framework was laid in 1987 at a workshop of the Canadian Agricultural Research Council, a consortium of researchers from industry, academia and government (Yarrow 2001). Agriculture and Agri-Food Canada, whose activities include encouraging industry to invest in research and development for new cultivars (Anstey 1986), was also responsible for regulating field experiments and approving cultivars for commercialization. This conflict of interest raised concerns, notably from the Auditor General, about the capacity of Agriculture and Agri-Food Canada to manage environmental risks independently (Yarrow 2001: 9). Such critics encouraged the creation of an autonomous regulatory agency, the CFIA. Although the capacity of the CFIA to regulate industry remains the object of concern (Royal Society of Canada 2001: 35–36; Auditor General of Canada 2004), Moore (2000) argues that the level of independence of the CFIA's regulators enables them to draw from international norms to a larger extent than American regulators. I have also argued elsewhere that the industry's connection to the Canadian state is mostly through Industry Canada, a ministry that promotes biotechnology, but which has no apparent influence on the CFIA's regulatory activities (Montpetit 2005). In short, the interconnection of the industry network to state actors is, in the worst scenario, moderate, which represents a slight difference with the network situation in the United States. As indicated in table 4.3, however, this nevertheless leaves the pro-technology network stronger than the network of actors with concerns about GMOs, a situation favorable to the adoption of permissive policies.

Table 4.3 summarizes the network situations in the areas of ART and GMOs. Pro-technology networks or policy communities of targets are strong in both areas, notably because of their cohesion. In contrast, networks of opponents or of actors with concerns about biotechnology are not quite as strong, especially in the area of GMOs where cohesion is low and interconnections with state actors loose. The women's network, however, is an exception. In fact, the recent cohesion and the interconnection of the network to the state offer a powerful explanation of the moderate rather than the permissive nature of the Canadian Assisted Human Reproduction Act, adopted in 2004.

## Country Patterns

The network situations presented in table 4.3 provide powerful explanations for the moderate ART policy design and the permissive GMO policy design. There remain, however, questions regarding these policy decisions. Two questions related to ART are best answered when examining the mix of power concentration and fragmented governance within the Canadian political system and one question, pertaining to GMOs, requires an examination of the international context: (1) Why did the government decide to adopt the Canadian Assisted Human Reproduction Act instead of opting for a nondecision as it did in the past given the brokerage nature of the Liberal Party of Canada, the election in the spring of 2004, and the possibility of passing the blame to provincial governments? (2) Why did the medical profession maintain its preference for self-regulation over time, in sharp contrast with the British medical profession? (3) What kind of pressure did the United States, Canada's powerful neighbor, exert in the area of GMOs? This last question will be examined in the last section of this chapter.

Canada, like the United Kingdom, has a Westminster type of parliamentary system enabling a single party to form the government. In contrast to the United Kingdom, however, Canada is a federal country in which provinces have substantial responsibilities, notably in the area of health care. These two elements of the Canadian polity were variously important for the ART and GMO policy designs.

As in the case of Britain, it is clear that the label of the governing party in Canada is a poor predictor of policy designs, and this may come as no surprise as the Liberal Party of Canada, in government since 1993, is known more as a brokerage party than an ideological party. In fact, the party's supporters form a coalition of actors with diverging interests (Clarke et al. 1996). In the area of ART, party politics nevertheless did matter. I have argued elsewhere that the Liberal Party of Canada's decision to abandon the adoption of the Human Reproductive and Genetic Technologies Bill in 1997 was motivated, in part, by upcoming elections (Montpetit 2004: 80). At the time, the Reform Party was the official opposition and was fully ready, along with the religious network, to bring the issue of ART on the terrain of abortion. Given that some of the Liberal Party's supporters were openly pro-life, adopting this first ART government bill in 1997 could have come with an electoral cost to the Party. Therefore, it was in the electoral interests of the Liberal Party to allow the Human Reproductive and Genetic Technologies Bill to die on Parliament's order paper, evoking a tight parliamentary schedule as a justification.

Interestingly enough, the Canadian Assisted Human Reproduction Act passed in the Senate just a few months before the Liberal government called an election in the spring of 2004. Just like in 1997, the issue was still a hot one, even among supporters of the Liberal party. Several liberal MPs, holding pro-life beliefs, even voted against the bill during the House's third reading in the fall of 2003. Between 1997 and 2004, however, party politics in Canada changed significantly. In fact, the Canadian Alliance (which had been formerly the Reform Party) merged with the Progressive Conservative Party to form a countrywide party, the Conservative Party of Canada. Unlike the Reform party, the Conservative Party sought a wide electoral basis in view of defeating the Liberal Party. In order to gain electoral support in Ontario and Québec, the two largest provinces, the Conservative Party had to distance itself from the religious network and, as much as the Liberal Party, avoid issues such as abortion. In other words, the replacement of the Reform Party by the Canadian Alliance and eventually by the Conservative Party gradually reduced the electoral risk to the Liberal Party of legislating on ART. Unlike in 1997, no party in Canada now has an interest in debating the abortion issue during electoral campaigns. In fact, the Canadian Assisted Human Reproduction Act did not haunt the Liberal Party during the electoral campaign of the spring of 2004. Without a party specifically representing Western Canada, brokerage should dominate party politics, leaving ART policy design in the hands of policy networks, just as was the case with the GMO policy design.

Federalism, in contrast to party politics, should have a lasting and consistent influence on policy designs in ART and to a lesser extent on GMO. The main impact of federalism in the area of ART, I argue, is on the preference of the actors of the medical network for self-regulation. In 1997, the Liberal government blamed the provinces for the failure of the Human Reproductive and Genetic Technologies Bill. Much of the opposition to the bill was related to the absence of provisions for the creation of a regulatory agency. The reason why a regulatory agency was absent, the federal government suggested, was because provinces had not given their agreement to what several of them viewed as an intrusion into their jurisdiction over health. In other words, provinces contributed to the veto in 1997 over the creation of a regulatory agency for ART. In maintaining a firm position in favor of self-regulation, the actors of the medical network acted in the belief that provinces would again prevent the adoption of a comprehensive act and therefore did not need to cooperate actively with federal policymakers. In contrast, medical actors in the United Kingdom proactively offered their cooperation in view of creating a state authority to license fertility clinics and approve ART research because they knew they were standing in the "shadow of hierarchy" (Scharpf 1997). It is better to cooperate in the hopes of influencing decisions, the shadow of hierarchy reasoning goes, than risk an opposition to a policy the government can in any case impose. In Canada, the actors of the medical network simply believed the federal government could not impose its policy because of the fragmentation of governance in this area. It should be underlined that it is still unclear whether the federal government had the authority to adopt the Canadian Assisted Human Reproduction Act without the consent of provinces. In fact, the Quebec government decided, in the fall of 2004, to challenge several sections of the Act in court.

The issue of GMOs is much less an object of intergovernmental struggle. Provincial governments recognize the authority of the federal government to manage the health and environmental risks associated with GMOs. However, provinces can decide, arguably, to ban GMOs on their territory or require labeling. At the time of writing, Prince Edward Island was holding public hearings on the possibility of banning GMOs from the province. Quebec's provincial government was elected on the pledge to make the labeling of GMOs mandatory and the agriculture parliamentary commission urged the government to make good on its promise in a widely distributed report published in 2004. Interestingly enough, the provincial Premier blames the province's own inaction thus far on labeling on the federal government, arguing that labeling only in Quebec would be ineffective and therefore that a federal policy is required. In any case, if policy change toward a more restrictive Canadian GMO policy is at all possible, it will likely come from the provincial channel.

To summarize, the women's network, in favor of a comprehensive federal policy, was fortunate that party politics returned to a regular competition between two brokerage-style political parties, after ten years of competition involving a social conservative party with a regional electoral base in Western Canada.[2] In 2004, no political party had an interest, as much as did the Reform Party in 1997, in running the electoral campaign on the morality of the federal ART policy. It was also fortunate for the women's network that the federal government chose to legislate despite provincial opposition. The courts will, however, decide whether or not the federal government had the legal authority to do so in the coming months or years. While federalism worked to prevent policy change in the area of ART, albeit ineffectively in 2004, it may encourage change in the area of GMOs in the near future, although any firm conclusion on this matter would be premature.

## Internationalization

Internationalization is not as relevant to the explanation of the Canadian ART policy design as it is to explain some limited aspects of the GMO design. As argued above, some forms of policy transfer from the United Kingdom occurred, even though the Canadian design ended up being more restrictive than the British ART policy design. In addition, the transfer did not occur through international institutions, but was realized in a direct fashion through exchanges between Canadian and British civil servants as well as between other important network actors.

In the GMO area, the influence of internationalization is not what common sense would suggest. Indeed, internationalization has not lowered the standards of environmental and health risk assessments to those of the United States, the main trading partner of Canada for agricultural commodities, but has enabled Canadian policy-makers to justify higher standards, even if they still appear permissive in comparison to those of European countries. As explained above, the interconnection of the industry network to the state is moderate. Industry Canada and Agriculture and Agri-Food Canada are the key state partners of the agri-food industry. The management of risks related to GMOs, however, is the responsibility of neither ministry, but is handled by the CFIA, an autonomous agency. The relative autonomy of the CFIA from industry and from the rest of government has enabled the agency's scientists to draw from international norms to a larger extent than scientists responsible for the regulation of GMOs in the United States Department of Agriculture, a department tightly interconnected to industry. CFIA scientists participated intensively in the development of international norms by the OECD, the World Health Organization, as well as the Food and Agriculture Organization presented in chapter 2. Around these three international organizations, an authentic epistemic community interested in the assessment of GMO's health and environmental risks was formed; several scientists involved in the Canadian regulatory processes were key players in this community. As Haas (1989) argues, such an epistemic community can contribute to aligning a country's policy on international norms. In the face of criticism regarding the use of substantial equivalence as a decision threshold, the Government of Canada replied that Health Canada's scientists contributed to the development of the international norm suggesting that familiarity and substantial equivalence should be used as a comparative tool only in a thorough risk assessment process and that the government abides by this norm. In contrast to the United

States, the government insists, no GM food or cultivar escaped risk assessment in Canada because of familiarity or substantial equivalence (Government of Canada 2001: 11).

In other words, the GMO network situation described in table 4.3 enabled the CFIA's scientists to participate in an international network, dominated by scientists and capable of developing norms that do not obviously
serve the interests of industry (Tiberghien and Starrs 2004; Coleman and
Gabler 2002). As suggested in chapter 1, actors somewhat concerned about
biotechnology successfully used international norms. However, they were
not sufficiently concerned about biotechnology to turn the Canadian GMO
policy into a restrictive one. Internationalization, therefore, had a small impact in comparison to countries where biotechnology opponents have a
greater capacity to use international rules.

It is of significant theoretical interest that it was the domestic network
conditions that enabled internationalization to influence GMO policy. What
mattered is not so much internationalization in itself but its combination
with a specific network situation. In the ART area, the domestic network
situation, characterized by a tight interconnection between Health Canada
and the women's network, did not enable the same extent of drawing upon
international norms. The combination of variables raised by this issue is discussed at length in the last chapter of this book.

## CONCLUSION

The Canadian ART policy is more restrictive than the Canadian GMO policy. Stem cell research, for example, cannot be conducted from embryos
created specifically for this purpose and must be licensed by a regulatory
agency. In contrast, developers of GMO cultivars simply notify the CFIA of
their field trials and abide by relatively permissive regulations. Between
1998 and 2003, the CFIA received 5,862 notifications of field experiments
on plants with novel traits (Auditor General of Canada 2004: 4.25). In fact,
the number of field trials and of authorizations for commercialization rose
much faster than the resources of the CFIA (Leiss and Tyshenko 2002:
326–27).

An examination of policy networks provides a straightforward explanation
for these policy differences. In the ART area, the network of actors demanding restrictions provided tighter connections to the state than the network of targets that preferred a permissive policy. In contrast, in the GMO
area, no network of actors demanding restrictions enjoyed a tight interconnection with state actors. Even though the industry network, forming a

community of targets, provided only a moderate interconnection, it sufficed to avoid a restrictive GMO policy design. These findings largely confirm the network hypotheses presented in chapter 1. This network explanation can be complemented by a discussion of party politics, concentrated and fragmented governance, all having distinctive implications for networks and biotechnology policy design. While party labels do not have the influence predicted in chapter 1, I have argued that the gradual transformation of the Reform Party from a regional party into a party seeking a countrywide electorate acted in favor of the women's network's demands for ART policy restrictions. While the concentration of governance, as suggested in chapter 1, appeared consistent with the permissive preferences of the community of targets in the GMO sector, fragmented governance did not so much provide access to actors concerned about biotechnology as it weakened the community of targets. This represents a small difference from the hypothesis, formulated in chapter 1, regarding fragmented governance. Factors associated with internationalization did not have a sufficient impact to alter, other than in a small way, the restrictiveness or the permissiveness of Canadian biotechnology policies in both sectors. In Canada, domestic networks combined with domestic institutions to shape biotechnology policy. In other words, if Canada played in the big leagues of the British and American players, it did not entirely lose its autonomy.

## NOTES

1. www.cwb.ca/en/topics/biotechnology/closing_gap.jsp; consulted April 14, 2005.
2. The return to a normal situation is not entirely clear since the Bloc québécois, a regional party, still easily obtains the majority Quebec's seats. However, this has no bearing on ART policy.

## REFERENCES

Anstey, T. H. 1986. *One Hundred Harvests: Research Branch, Agriculture Canada, 1886–1986*. Ottawa: Research Branch, Agriculture Canada.

Auditor General of Canada. 2004. *Canadian Food Inspection Agency: Regulation of Plants with Novel Traits*. Ottawa: Office of the Auditor General of Canada.

Barrett, K., and E. Abergel. 2000. Breeding Familiarity: Environmental Risk Assessment for Genetically Engineered Crops in Canada. *Science and Public Policy* 27 (1): 2–12.

Canadian Food Inspection Agency. 2004. Long Term Testing/Substantial Equivalence. www.inspection.gc.ca/english/sci/biotech/reg/equive.shtml. (accessed July 14th, 2005).

Clarke, H. D., J. Jenson, L. LeDuc, and J. H. Pammett. 1996. *Absent Mandate: Canadian Electoral Politics in an Era of Restructuring*. Vancouver: Gage Education Publishing Company.

Coleman, W. D., and M. Gabler. 2002. Agricultural Biotechnology and Regime Formation: A Constructivist Assessment of the Prospects. *International Studies Quarterly* 46 (4): 481–506.

Doern, B. G., and T. Conway. 1994. *The Greening of Canada: Federal Institutions and Decisions*. Toronto: University of Toronto Press.

Government of Canada. 2001. *Action Plan of the Government of Canada in Response to the Royal Society of Canada Expert Panel Report*. Ottawa: Department of Public Works and Government Services.

Haas, P. M. 1989. Do Regimes Matter? Epistemic Communities and Mediterranean Pollution Control. *International Organization* 43 (3): 377–403.

———. 1992. Introduction: Epistemic Communities and International Policy Coordination. *International Organization* 46 (1): 1–36.

Jasanoff, S. 1995. Product, Process or Program: Three Cultures and the Regulation of Biotechnology. In *Resistance to New Technology*. Ed. M. Bauer. Pp. 311–31. Cambridge: Cambridge University Press.

Leiss, W., and M. Tyshenko. 2002. Some Aspects of the 'New Biotechnology' and its Regulation in Canada. In *Canadian Environmental Policy: Context and Cases*. Eds. D. L. Vannijnatten and R. Boardman. Pp. 321–44. Oxford: Oxford University Press.

Montpetit, É. 2003. *Misplaced Distrust: Policy Networks and the Environment in France, the United States and Canada*. Vancouver: UBC Press.

———. 2005. A Policy Network Explanation of Biotechnology Policy Differences between the United States and Canada. *Journal of Public Policy* 25 (3): 339–66.

Montpetit, É., F. Scala, and I. Fortier. 2004. The Paradox of Deliberative Democracy: The National Action Committee on the Status of Women and Canada's Policy on Reproductive Technology. *Policy Sciences* 37 (2): 137-57.

Moore, E. 2000. Food Safety, Labelling, and the Role of Science: Regulating Genetically Engineered Food Crops in Canada and the United States. Paper for the ECPR Joint Sessions Workshop on the Politics of Food, Copenhagen.

Mulkay, M. 1997. *The Embryo Research Debate: Science and the Politics of Reproduction*. Cambridge: Cambridge University Press.

Royal Society of Canada. 2001. *Elements of Precaution: Recommendations for the Regulation of Food Biotechnology in Canada*. Ottawa: Royal Society of Canada.

Scharpf, F. W. 1997. *Games Real Actors Play: Actor-Centered Institutionalism in Policy Research*. Boulder: Westview Press.

Tiberghien, Y., and S. Starrs. 2004. The EU as Global Trouble-Maker in Chief: A Political Analysis of EU Regulations and EU Global Leadership in the Field of Genetically Modified Organisms. Prepared for the 2004 Conference of Europeanists, the Council of European Studies.

Yarrow, S. 2001. Environmental Assessment of the Products of Biotechnology. Proceedings of *Food of the Future: Genetically Modified Foods*, Simon Fraser University.

# 5

# A CONTRAST OF TWO SECTORS IN THE BRITISH KNOWLEDGE ECONOMY

*Éric Montpetit (Université de Montréal)*

The United Kingdom has the reputation of being a leading country in biotechnology. The first IVF baby was born in the UK in 1978 and the first animal cloning was carried out in the UK. Not all scientists, however, would agree that the UK provides the ideal regulatory environment for conducting biotechnology research or commercializing biotechnological products. In fact, in this chapter I show that sharp differences exist between ART and GMOs. While the ART policy design is permissive, the GMO design is restrictive.

This sharp policy difference is best explained in large part, I argue, by differences in policy networks. Because the UK's ART policy was designed in the 1980s, which is relatively early in the history of biotechnological development, the characteristics of the ART network were relatively simple, involving few organized groups that were effective in opposing a permissive policy. This network situation helped the prime minister manage the policy process so as to provide organized groups of biotechnology supporters with sufficient time to make Conservative Members of Parliament (MPs) increasingly favorable to a permissive policy. In contrast, the GMO network situation was far more complex, including several autonomous regulatory agencies over which the prime minister had little control. With the backing of the European Union, the network of environmentally concerned actors was able to exert significant influence during GMO policy development at the end of the 1990s. The chapter first presents the differences between the permissive ART policy and the restrictive GMO policy and then moves on to a discussion of policy networks, the political architecture of the British state and Europeanization as potential explanatory factors.

## PERMISSIVE ART POLICY

Of all the countries examined in this book, the UK and Germany were the only ones to have designed major ART policies in the 1980s. In both countries, the policy debates of the 1980s focused primarily on the question of whether embryos had rights in the same way as humans. In contrast to Germany's Embryo Protection Act of 1990, however, the UK's 1990 Human Fertility and Embryology Act provided far more than a clarification of the rights of embryos and their implication for clinicians. The act also set up a licensing agency, covering both clinics and research laboratories, which has been relatively permissive at granting authorizations for technologies that are prohibited in most countries.

The key ideas behind the Human Fertility and Embryology Act came from the Warnock Committee, which reported to the UK parliament in 1984, and which was the world's first major inquiry conducted on ART. The Warnock Committee was set up in reaction to the birth of the first IVF baby in 1978, and therefore the committee's work was largely centered on research relating to IVF and its clinical applications. The inquiry contributed to the framing of ART policy debates in several countries, notably Canada and France, although the development of new technologies has widened considerably policy debates in those countries. In comparison to other countries, the UK was developing its ART policy in a context of relative technological simplicity. The use of technologies such as embryonic stem cells to cure terrible diseases or prenatal diagnosis (also known as preimplantation diagnosis) to enable the selection of embryos free from genes of diseases carried by parents were considered "unlikely feasible . . . for some considerable time" (Warnock 1984: 75). In the 1980s IVF was really the only technology at stake as embryo research was aimed primarily at improving the success rate of this reproductive technology. In comparison with today, the range of issues under consideration for the development of an ART policy design was narrow. Nevertheless, the Warnock Committee did not ignore possible future developments in ART, devoting a full chapter of its report to this matter, and certainly accepted "that scientists must not be unduly restricted in pursuing their research interests especially when this may produce direct therapeutic benefits" (Warnock 1984: 75). To avoid inhibiting future developments in ART, while ensuring the public's protection, the Warnock Committee believed a "statutory licensing authority" was the best policy instrument.

The Human Fertilisation and Embryology Act of 1990, which remains the main piece of legislation governing ART in the UK, closely followed the recommendations of the Warnock Committee. Although the act prohibited extreme practices such as sex selection, it permitted embryo research. This Act created the Human Fertility and Embryology Authority (HFEA), a nondepartmental agency responsible for the delivery of licenses to clinics offering fertility treatments and specific human embryo research projects. Originally, the act specified five purposes for which the HFEA could license embryo research projects: improving fertility treatments, looking into the cause of miscarriage, looking into the cause of congenital disease, developing new approaches to contraception, and improving the early diagnosis of genetic disease.

Technological developments considered unfeasible in the 1980s appeared within reach in the 1990s. Stem cell research notably brought new ideas and new actors into the policy debates. Infertile couples were no longer the only beneficiaries of ART, victims of genetic diseases suddenly had much to gain from human embryo research. In light of these developments, British policy-makers felt it necessary to amend the Human Fertility and Embryology Act in 2001. The amendment added the purposes of increasing knowledge about embryos, serious genetic diseases, and their treatment to the five purposes under which the HFEA can authorize embryo research projects. Concretely, this amendment empowered the HFEA to authorize research projects on embryonic stem cells, including stem cells from cloned embryos. As of March 2005, two projects employing therapeutic cloning had been approved by the HFEA. As far as research is concerned, the UK's ART policy is indisputably permissive, despite HFEA's strict licensing conditions, monitoring of research and transparency. In comparison, the permissiveness of the American ART policy is the result of an absence of public oversight on private research.

As mentioned above, the HFEA also licenses clinics that offer fertility treatments. On this particular matter, the UK's approach is more cautious, in fact qualifying as an intermediate policy design. Indeed, the act encourages the HFEA to adopt the perspective of patients using these services, implying that regulations should ensure the safety of patients and their unborn children without unreasonably restricting access. For instance, the act specifies that gametes should not be given in exchange for money, "unless authorised by directions." Worried about the availability of gametes, the HFEA decided that payments of a maximum of £15, plus reasonable expenses, could be made in exchange for gametes. The identity of gamete donors is protected by the act, although information on their genetic background is collected by the HFEA and can be obtained by patients. In terms of prenatal diagnosis (PND) and tissue typing, the act can also be characterized as cautious. Tissue typing consists of conceiving a child who, thanks to a PND, possesses a genetic makeup consistent with the tissue requirements to cure a child afflicted by a specific genetic disease. Instead of adopting a permissive or a restrictive approach to these technologies at the outset, the HFEA has studied closely each case, trying to assess the risks and benefits of these practices for the patient and children. For example, the HFEA wanted patients to avoid undergoing any PND whose efficacy remained unclear. As knowledge has accumulated on PND, the HFEA began granting licenses to clinics to offer PND for specific diseases. Eight clinics have received these licenses in the UK and the HFEA is currently streamlining its process for granting new licenses related to diseases for which particular diagnoses have clearly been shown to be efficient.

A license to employ tissue typing techniques was granted only once by the HFEA. The demand originated from the Hashmi family, who wanted to use PND to conceive a child whose tissues would be compatible with those of their six-year-old son who needed a bone marrow transplant. Opponents to this technology doubted that the HFEA possessed the legal authority to authorize tissue typing and engaged in a long legal battle. The issue was finally settled in the spring of 2005 when the Law Lords, the highest court in the UK, confirmed the authority of the HFEA to authorize tissue typing. The HFEA has exercised this authority with caution, examining each individual case on its merits.

Lastly, the UK's policy does not set any specification on the marital status of patients, their sexual orientation or their age. On these matters, the policy is clearly permissive. In fact, the National Health Service generally insures licensed treatments, although funding these treatments has been difficult in areas experiencing financial difficulties (Blank 2004: 133–34). The UK's ART policy design is summarized in table 5.1.

## RESTRICTIVE GMO POLICY

The contrast between the UK's ART and GMO policies is sharp. While the UK is one of the few places in the world where therapeutic cloning is per-

**Table 5.1.   Current UK ART Policy Design**

|  | Research | Practice of fertility treatments | Access for patients |
|---|---|---|---|
| Policy Designs | - Creation of embryos for research and therapeutic cloning permitted | - The HFEA has set a maximum of £15, plus reasonable expenses for any gamete donation<br>- Gamete donations have to be anonymous<br>- Prenatal diagnosis limited to predefined diseases<br>- Tissue typing examined on a case-by-case basis | - Access unrestricted |
| Type | Permissive | Intermediate | Permissive |

mitted and where research on cloned embryos is underway, researchers face difficulties obtaining permission to experiment on GMOs in field trials. If GMO research is difficult, the commercialization of GMOs is largely impossible, and most cultivars and novel foods are still awaiting approval by the responsible authorities.

As Toke and Marsh (2003) argue, the UK's GMO policy has not always been this restrictive. In the 1980s and for the major part of the 1990s, the Ministry of Agriculture, Fisheries and Food (MAFF), and the biotechnology industry in general, dominated policy-making in this area. Prior to 1998, for instance, the Advisory Committee on Releases into the Environment (ACRE), the scientific body that assesses the risks of GMOs to be released into the environment, was dominated by scientists who had links to the biotechnology industry (Toke and Marsh 2003: 236). Under these conditions, biotechnology R&D spending was high in comparison to other European countries (Levidow and Carr 2000: 6). Only after a reform of its regulatory framework did the UK policy design become restrictive.

As with all EU member states, the UK's regulatory framework is the result of the translation of Directive 90/220/EEC into national law, with slight changes resulting from the more recent Directive 2001/18. The key features of the English system are contained in Part VI of the environmental Protection Act of 1990 and in the Genetically Modified Organisms Regulations of 1992 (Northern Ireland, Scotland and Wales have their own acts and regulations). ART research and practices are authorized by licenses issued by the HFEA, which represents a delegation of authority to an independent agency. In contrast, the government directly gives authorizations for GMO field trials and commercialization in the UK. Applications for

field trials, so-called part B releases in the EU directive, have to be submitted to the Joint Regulatory Authority, which advises the secretary of
state at the Department for Environment, Food and Rural Affairs (DE-
FRA). Complete applications are then forwarded to ACRE, which under-
went a reform in 1999 that I discuss below. Any application regarding a
product likely to enter the food chain is also forwarded to the Advisory
Committee on Novel Foods and Processes (ACNFP) and the Advisory
Committee on Animal Feedingstuffs (ACAF), which both advise the Food
Standards Agency, created in 2000. Commercial uses of GMOs, or the so-
called part C releases in the European directive, are also channeled
through this system. Naturally, any GM food and cultivars commercialized
in the UK are subject to the traceability and labeling regulations of the EU.

The key policy changes that moved the country in the direction of greater
restrictions were made between 1998 and 2000. First, in 1998, the minister
of the environment called for a halt in the commercialization of GMOs, just
before the completion of part C approvals for some GM cultivars in the UK.
Negotiations were then undertaken with the Supply Chain Initiative on
Modified Crops (SCIMAC), a group representing the farming and biotechnology industry, over a program of farm-scale trials of GMOs to assess their
environmental impact. The farm-scale trials anticipated the European Directive 2001/18, which required a thorough environmental assessment of
GMOs. The agreed upon farm-scale evaluations program was carried out
on three herbicide tolerant cultivars between the spring of 2000 and the
winter of 2003 and was funded entirely by the government. A committee of
independent scientists, the Scientific Steering Committee (SSC), conducted the evaluations. Although rather positive about the effects of GM
maize on biodiversity, the final report of the SSC suggests caution given the
difficulty in predicting the long-term effects of GMOs (UK farm-scale
Evaluations Research Team and the Scientific Steering Committee, 2003).
Certainly, the SSC set new and restrictive standards for the environmental
risks posed by GMOs. Second, in 1999, ten of the thirteen members of
ACRE were replaced by scientists who had no link with the biotechnology
industry (Toke and Marsh 2003: 237). Moreover, the terms of reference of
the advisory committee were expanded to include thorough assessments of
"ecological imbalance," which might arise from the cultivation of GM plants
(Levidow and Carr 2000: 4). Previously, ACRE's assessments were mostly
related to the agronomic potential and risk of GMOs (Toke and Marsh
2003: 236; Levidow and Carr 2000: 3). Third, the Labour government created the Agriculture and Environment Biotechnology Commission (AEBC)
in 2000 to stimulate a public discussion and to independently advise the

**Table 5.2.   Current UK GMO Policy Design**

|  | Field Trials | Commercialization and Cultivation | Consumer and environmental protections |
|---|---|---|---|
| Policy Design | - Field experiments allowed on a case-by-case basis | - Sales de facto prohibited | - Labeling requirements defined at the European level<br>- Traceability requirements defined at the European level |
| Type | Intermediate | Restrictive | Restrictive |

government on the broad concerns generated by GMOs. The AEBC, notably through its interventions during the farm-scale evaluations, encouraged caution with respect to GMOs (UK Agriculture and Environment Biotechnology Commission 2001). The FSA was similarly mandated to stimulate reflection on the health safety issues related to GM foods (UK Food Standards Agency 2003). Fourth, the Labour government implemented transparency measures in the treatment of applications for GMO releases and commercialization. DEFRA's Joint Regulatory Authority, as well as the FSA, now publishes all the authorization files on the Internet.

These measures make for a restrictive policy design. Not even the maize shown to have positive effects on biodiversity by the farm-scale evaluations was approved for commercialization. Since October 2002, 182 applications were filed in the EU under part B of Directive 2001/18. Of these, only four were filed with the UK's Joint Regulatory Authority (Biotechnology & GMOs Information Website). Since January 2003, only four of the twenty-eight applications filed in the EU under part C were filed in the UK. The industry rightly believes that the competent authorities of other EU countries are more permissive than the UK's (see table 5.2).

## EXPLAINING THE DIFFERENCES BETWEEN ART AND GMO POLICY DESIGNS

At first glance, the aforementioned policy differences are puzzling. Why would a country decide to restrict one key area of the knowledge economy, but not another one? Furthermore, the area selected for restrictions, GMOs, is precisely the one that has the largest economic potential in the foreseeable future. As explained below, significant policy-making activities in both areas occurred at different periods. While the permissive ART policy was developed in the 1980s and remained largely unchanged, the GMO policy, which was initially permissive, only became restrictive at the end of the 1990s. A full decade separated significant policy-making activities in these two areas and this decade was one of political change. The Conservative Party, which had formed every government since 1979, was defeated in 1997 by the Labour Party. In addition, and as indicated in chapter 2, European integration accelerated significantly during this decade. Finally, mad cow disease, a major food safety scandal, increased the public's awareness of potential failures of science-based regulatory processes.

## Policy Networks

The GMO and ART networks also reveal sharp contrasts. In the ART sector, opponents reacted effectively to the report of the Warnock Committee in 1984, quickly obtaining support for a restrictive bill. However, unconvinced by the necessity of this bill, the British government bought time to enable scientists, the main target group of this bill, to form a cohesive policy community. This policy community worked hard to convince members of Parliament and public opinion that the adoption of a restrictive act was unwise. The policy community effectively undermined the cohesion of the network of opponents and deprived it of an interconnection with decisive state actors. These were the attributes that had made the network of opponents so powerful in the aftermath of the publication of the Warnock Committee report. To this day, the policy community of pro-technology actors has remained powerful enough to avoid restrictive policies, while the network of ART-concerned actors has formed a weak issue network, having failed to recreate the cohesion and interconnection.

In the GMO sector, the situation is reversed. A cohesive and closely interconnected community of pro-technology actors dominated this sector in the 1980s and early 1990s. Technology-concerned actors contributed to breaking the cohesion of this community in the mid-1990s and developed close interconnections with powerful state actors. The pro-technology network was unable to avoid becoming the main target of restrictive policies adopted at the end of the 1990s.

One could argue that GMOs and ARTs are so different in substance that the policy differences are unsurprising. I argue, however, that the differences in the technological and political contexts between the 1980s and the 1990s were more significant for network formation and eventual policy decisions than the substance of both policy areas. A rapid comparison with

Canada may be useful here. Canada's ART policy, developed between 1993 and 2004, was inspired by the UK's Human Fertilisation and Embryology Act of 1990. In terms of restrictiveness, however, the Canadian policy has much more in common with the UK's GMO policy, whose development also took place primarily towards the end of the 1990s. At the end of the 1990s, that is after stem cell research and other technological developments had become ART policy issues, ART-concerned actors in Canada formed a wide network, much like the British network of GMO-concerned actors during the same period. And again like the British network of GMO-concerned actors, the network of ART-concerned actors in Canada was also closely interconnected with decisive state actors, more so than the community of target groups. In short, the context of the 1980s versus the context of the 1990s has more bearing for network formation than timeless differences in the substance of GMOs and ART.

Indeed, the network situation that prevailed in the UK in the 1980s in the area of ART was simple. The Warnock Committee report suggested that embryos developed gradually into humans and that therefore they should not be granted the same rights as human beings during their first fourteen days of life. This fourteen-day limit, which became an international standard adopted by several countries, was justified by the absence of a central nervous system that would enable embryos to feel pain prior to that point (Warnock 1984: 65). As several groups firmly believed that embryos deserved the same rights as human beings from the moment of conception, networks of what would become the key actors during ART policy development in the UK formed around this controversial aspect of the Warnock report.

Two networks dominated policy debates over the development of the Human Fertilisation and Embryology Act: the religious network and the medical network. The Society for the Protection of Unborn Children (SPUC) is the main interest group of the religious network. SPUC was created in 1967 after the founding members realized that people supporting the rights of unborn children were poorly organized to exert influence on the Medical Termination of Pregnancy bill of 1966, which liberalized abortion in the UK. SPUC reacted strongly to the report of the Warnock Committee. Although officially a nonreligious group, SPUC's anti-abortion position was vastly supported among the members of the main religious confessions in the UK. In fact, when the Warnock Committee released its report, SPUC could count on the main church representatives to convey its criticisms to the public and especially to Parliament. The House of Lords had several members with close ties to churches who argued that embryo research went against the sanctity of life (Mulkay 1995a). And the reactions of the religious network were initially effective, prompting the introduction of a private member's bill by Enoch Powell, the Unborn Children (Protection) Bill, in February of 1985. The bill was a clear rejection of the recommendations of the Warnock Committee, requiring immediate implantation of embryos, thereby prohibiting embryo research. Moreover, parliamentarians warmly welcomed the bill with 238 MPs supporting it on its second reading and only sixty-six voting against it (Mulkay 1995b: 33). If the third reading was never completed, it was not for a lack of support, but it was due to a filibuster supported by a minority of MPs (Gunning and English 1993: 43).

This initial strong and effective reaction by the religious network served to consolidate the medical community in support of the Warnock report. The key actors of the medical network were the Medical Research Council (MRC), the Royal College of Obstetricians and Gynaecologists (RCOG), PROGRESS, an organization of scientists formed in reaction to the Powell bill, and the Royal Society. The reaction of the religious network to the Warnock report left the impression that IVF and related research could lead to terrible abuses on the part of the medical community and much of the support for the Powell bill related to this belief. In order to dispel fears related to ARTs, but also to prevent the adoption of a law targeting their members with restrictions, the MRC and the RCOG decided to create a self-regulating body, the Voluntary Licensing Authority (VLA). The VLA's purpose was to show that in absence of state regulations, the medical community was assuming its responsibilities by establishing norms of acceptable practices for clinics and laboratories (Gunning and English 1993: chapter 4). Meanwhile, PROGRESS began, at once, a campaign to convince the public and MPs of the benefits of ARTs. Personal stories of patients were widely diffused in the media and MPs were invited to visit fertility clinics and laboratories. As Mulkay (1995b: 49) argues, this campaign was largely successful. The treatment of ART in the press became positive and opinion polls turned increasingly in favor of embryo research. By 1988, after the release of the government's white paper on ART, a majority of debaters in Parliament were in favor of ART, a sharp contrast with the parliamentary debates over the Powell bill just three years prior. In 1989, Enoch Powell's private member bill was reintroduced in the House of Lords, but this time received little support. To illustrate the shift in support for ART in Parliament Mulkay (1995b: 34) writes that "24% of those MPs who voted against embryo research in the Commons in 1985 voted for it in 1990, while all those who voted for embryo research in 1985 did so once again at the later date."

This opinion change in the House of Commons reflected the change that the campaign of the medical network had on the religious network. SPUC counted on the main religious confessions of England to exert policy influence. In contrast to SPUC, however, the members of the main religious confessions did not have a dogmatic position on ART. In an enlightening article, Mulkay (1995a) shows the extent to which church leaders gradually became divided over the status that should be granted to embryos during the 1980s. Several members of the Church of England, having been exposed to the arguments of the medical community, began to believe that embryo research, with its potential to cure terrible diseases, could also be justified in the name of the sanctity of life. In short, the cohesion of the religious network around the idea of an outright ban of embryo research gradually eroded. In fact, the recommendation of the Warnock Committee to regulate ART and the permissive policy trajectory established by the VLA had gained favor within the religious network by the end of the 1980s.

The ART network situation has changed very little since the adoption of the Human Fertilisation and Embryology Act in 1990, despite technological developments and a changing political context. The HFEA, created by the act, has become a powerful actor, in fact institutionalizing the policy orientations contained in the Warnock report. The HFEA is closely interconnected with the policy community of pro-technology actors, especially researchers, and therefore provides them access to the state. In contrast, the network of technology-concerned actors was never able to rebuild the interconnection it had to the state through Parliament in the mid-1980s. An inventory of consumer, disabled and pro-life groups, prepared for the Association of Medical Charities, Genetic Interest Group and Progress Educational Trust in 2003, reveals that most of the groups critical of human genetics and embryology not only lack resources, but also are rarely consulted by policy-makers (Festing, Gillott, and Tizzard 2003). In fact, some of these groups resort increasingly to courts in hope of influencing policy, but thus far with little success. The best-known example of a group resorting to this strategy is Comment on Reproductive Ethics (CORE), a group created in 1994, which basically shares the same perspective as SPUC on embryo rights. CORE, however, spends a large share of its resources on court cases. In 2001, the group filed a case against the HFEA, contesting the agency's legal authority to permit tissue typing. In a first court decision in 2002, CORE won a decision against the HFEA, raising hopes that the strategy was producing results. Hopes were rapidly deflated after CORE lost a first appeal in 2003 and the final appeal before the Law Lords in 2005. After the

enactment of a permissive ART policy in 1990, the changes in the technological and political contexts that occurred in the 1990s were insufficient to
change the network situation in a manner that would bring about restrictive
policy decisions.

The network situation regarding GMOs presents meaningful contrasts with
that prevailing for the ART policy design. First, actors vehemently opposed to
GMOs, in contrast to those in the religious network opposed to ART in the
mid-1980s, have always been disconnected from centers of policy decision. As
Toke and Marsh (2003: 235) argue, groups such as Greenpeace and Friends
of the Earth are all simply excluded from GMO policy-making networks.
Likely policy targets in the GMO sector, therefore, never feared facing restrictions to the extent that scientists did in the ART sector after the Powell bill
was introduced into Parliament in 1985. Therefore, the incentives for the
likely target groups in the GMO sector to join into a cohesive and tightly interconnected policy community were not as large as in the ART sector.

As in the ART network situation, two networks are important to explain
GMO policy design: an industry network, which like the medical network is
favorable to technology; and an environmental protection network, which
like the religious network is concerned with the consequences of technological advances. However, and here rests a second contrast between ART
and GMOs, some members of the industry network began distancing themselves from the permissive position that prevailed in the network before
1998. It should be recalled that the medical network remained cohesive
during the entire period of ART policy development and remains so today.
The industry network has a formal structure in SCIMAC, which coordi-
nates the activities of farm input suppliers, the main UK farm groups,
processors and distributors. For processors and distributors, the mad cow
crisis and its spillover effect on public opinion for all matters related to food
safety, including GMOs, was not unsubstantial. As Bernauer and Meins
(2003: 660) explained, the decision of Nestlé and Unilever to label their
products containing GMOs turned against them after activists began targeting these labeled products in consumer-oriented campaigns. In turn, the
success of these campaigns prompted several food chains to withdraw all
products containing GMOs from their shelves (Levidow and Carr 2000: 6).
From that point on, processors and distributors, in contrast to input suppliers, began supporting policies aimed at building consumers' confidence
rather than policies to encourage technological advances, hence the appeal
of the farm-scale evaluations to the former groups.

Third, moderate environmental groups, notably English Nature, Royal
Society for the Protection of Birds and the Game Conservancy Trust,

agreed on a cautious approach, involving risk assessments conducted by independent scientists. Again, agreement within the religious network around an appropriate policy for ART eroded over time, weakening the position of the network of actors concerned about technology in the area of ART. The policy preference of the environmental protection network found echo in the state, notably in the Department for the Environment, Transport and Region (DETR) under Meacher, a minister notoriously concerned about GMOs (Toke and Marsh 2003: 236–37). Simultaneously, the permissive approach lost ground within the state with the demise of the Ministry of Agriculture, Fisheries and Food (MAFF), which merged within DEFRA in 2001. The reform of ACRE, the creation of the SSC and of AEBC eventually provided additional support to the cautious preferences of the environmental protection network.

Table 5.3 shows the extent of the contrast between the ART and GMO network situations. Policy networks exert influence when they are cohesive and provide a tight interconnection with the state (Montpetit 2004). In the case of ART, the medical network, which comprised actors holding permissive policy preferences in a cohesive manner, provided tight interconnection with powerful state actors. I will elaborate further on this last point in the next section. In contrast, between 1984 and 1990, the religious network lost much of its cohesion. As representatives of the main religious confessions in Parliament provided the interconnection with the state, the efficacy of the interconnection eroded with the cohesion among religious leaders. In the area of GMOs, it is the pro-technology network's position that decreased in cohesion. In the face of increasing public resistance toward GMOs, interests diverged between input-providers on the one hand and processor as well as distributors on the other hand. In addition, the reorganization of the state through the creation of DEFRA, which reflected a change in approach from agricultural production toward rural sustainable development, contributed to a strengthening of the connection to the state of the network of technologically concerned actors in this area.

**Table 5.3.   Contrasting Network Situations between ART and GMO**

|  | Pro-technology networks | | Networks of actors concerned about technology | |
|---|---|---|---|---|
|  | Medical network (ART) | Industry network (GMO) | Religious network (ART) | Environmental network (GMO) |
| Cohesion | High | Low | Low | High |
| Interconnectedness | Tight | Loose | Loose | Tight |

## Country Patterns

Do these contrasts in network situations suffice to explain the differences in the ART and GMO policy designs? While policy networks provide the main explanation for the difference, it will be evident in this section that complementary factors also exerted an influence. Networks do not evolve in a vacuum, but rather are shaped by the context in which they are embedded (Marsh and Smith 2000), a point that is confirmed by the above discussion of the technological context. I will argue in this section that the architecture of the state also contributes to shaping networks in ways that also have policy implications. In the ART sector for instance, the concentration of power afforded by the British Westminster system of government was valuable to the policy community of target scientists in the 1980s. In the next section, I will further argue that European integration had the opposite effect (i.e., favoring the network of technology-concerned actors) in the GMO sector in the 1990s.

There are two related elements in the traditional architecture of the British state that could bear on biotechnology policy choices. The first element is the party government system. In the UK a single party always governs. Between 1979 and 1997, the Conservative Party formed the government. Since 1997, the Labour Party of Tony Blair has governed. The second element is the centralized structure of the traditional British state, known as the Whitehall model. Power in the UK has traditionally been concentrated in the hands of a prime minister, assisted by a professional bureaucracy. When the prime minister's party has a majority of seats in the House of Commons, which has been the case since 1979, she or he can control the legislative agenda, notably through party discipline. In other words, there are very few veto points in the way of the prime minister's preferred policy designs, which are carefully formulated by professional civil servants. There are debates as to whether this Whitehall model still accurately describes the functioning of the British state, as administrative reforms put in place since 1987 and throughout the 1990s have created several autonomous agencies, over which the prime minister appeared to have surrendered control (see Marsh, Richards, and Smith 2001; Pollitt and Bouckaert 2000). The comparison between ART policy development in the 1980s and GMO policy development in the 1990s provides some evidence of a change in the Whitehall model.

The importance of the concentration of power inherent in the Whitehall model is illustrated in the case of ART. I explained above that Enoch Powell's private member bill received wide support in Parliament when it was introduced. The support came massively from the Conservative Party with

89 percent of its MPs voting for a complete ban of embryo research in 1985. This vote among Conservative MPs should not be surprising, as the party has always emphasized established values, hence the suspicion against tampering with human life that ART permits. What is more surprising is that the Conservative government of Margaret Thatcher allowed a filibuster to kill Powell's bill. Indeed, the government had the power to extend the parliamentary session to allow the third reading of the bill. According to Mulkay (1995b: 39–43), members of Thatcher's government disagreed with the majority of Conservative MPs on embryo research. Thatcher and her various health ministers were not traditional conservatives, but progressive conservatives who agreed with the Warnock Committee that regulation of embryo research could be accomplished in a manner respectful of human life and still unleash enormous benefits. Instead of enabling the adoption of Powell's bill, the Thatcher government preferred to undertake consultations on three proposals: self-regulation as achieved by the VLA, regulated research and clinical applications as proposed by the Warnock Committee, or the complete ban of embryo research contained in the Powell bill. As discussed at length in chapter 4, the concentration of power did cast a shadow of hierarchy on the medical profession, contributing to its alignment behind the report of the Warnock Committee rather than behind the self-regulatory option (Montpetit, Rothmayr, and Varone 2005: 137–38). In any case, the consultation bought precious time for the campaign launched by the medical network and ended with the publication of a White paper that unsurprisingly supported the recommendations of the Warnock Committee. To pass the Human Fertilisation and Embryology Act, the Thatcher government did not even have to use party discipline: her informal veto of the Powell bill and the time the consultations afforded to the medical network were sufficient to turn public opinion and more importantly the opinion of several Conservative MPs. In short, if the Conservative Party label offers a poor explanation of the UK's permissive ART policy design, the concentration of power in the hands of the prime minister contributed significantly to the network situation and the policy design.

As suggested above, the concentration of power that characterized the 1980s, which is expected in the Whitehall model, might have diminished in the 1990s. The case of GMO supports this argument. Members of the Labour Party have been more optimistic about technological progress than traditional conservatives (Mulkay 1995b: 43). Prime Minister Tony Blair is particularly enthusiastic for the hopes that scientific progress and new biotechnologies represent. Blair's optimism appears in very complimentary comments that he made with respect to a report of the Nuffield Council on

Bioethics (1999), which insists on the advantages of genetic engineering. For Blair, there is no doubt that GMOs offer more advantages than disadvantages, and that science will manage to overcome the reservations of a rather hostile British public. Blair's determination, however, did not suffice to stop GMO policy developing more restrictions.

Three interrelated differences with the ART case might have contributed to this situation. First, in the case of ART, Thatcher's views were in line with her ministers, notably her health ministers and secretaries, and, more generally, with the civil service. In contrast, Blair's environment minister, Michael Meacher, did not share his prime minister's optimism about GMOs. This situation bears similarities with Germany, where the chancellor's optimism about biotechnology was not shared by his environment minister. In any case, the environmental protection network in the UK found in Meacher a governmental ally. Indeed, it was Meacher who decided to launch the farm-scale evaluations and who broadened the terms of reference of ACRE to include assessments of the impact of GMOs on biodiversity (Levidow and Carr 2000: 4).

Second, Blair's optimism about biotechnology was matched by his faith in the capacity of science to prove the safety of GMOs. Therefore, he did not resist pressures to increase the presence of independent scientists on regulatory devices. For example, the composition of ACRE, which included independent experts, was institutionalized by Blair, thanks to the Nolan principle of the cabinet office, which aims at preventing any political interference during recruitment (Toke and Marsh 2003: 236). Also, the new Food Standards Agency (FSA), within which the ACFNP operates, is a nonministerial agency, thus guaranteeing a broad independence vis-à-vis the government. AEBC is yet another such autonomous organization created by the Labour government. As affirmed by an interview subject, "we owe to the government the creation of these independent organizations and it is him which pays our wages. However, it does not have any influence on the organization's reports and the decisions." It should be reminded that in the ART sector, a single autonomous organization, the HFEA, was created in 1990. What is more, this organization evolved from the VLA, a self-regulatory body, which itself evolved from the Warnock report supported by the prime minister of the time. In contrast, the backgrounds of the numerous autonomous agencies of the GMO sector are diverse and they thereby provide access points to the state to diverse actors, including the members of the environmental protection network who do not share Blair's optimistic policy preference. In other words, Blair's control of GMO policymaking in the 1990s was weaker than Thatcher's control of ART policy-making in the 1980s, indicating a possible significant change in the White-hall model.

## Europeanization

I have just argued that the concentration of powers in the hands of the British prime minister had important bearings on the ART network situation and policy design, but not in the case of GMOs. The situation is reversed as far as globalization and Europeanization are concerned. As explained above, the UK was an early designer as far as ART is concerned. ART policy development, for the most part, took place in the 1980s, and the UK's comprehensive act on this matter was adopted in 1990, a time when most countries examined in this book were just beginning to consider their own ART policy. As indicated in other chapters, the UK is in fact the point of departure of policy transfers, transiting internationally and ending up in several countries, notably Canada and France.

In contrast, the UK's GMO policy was made more restrictive in an international and especially European environment that provided both inspiration and specific directives. In fact, the restrictive labeling and traceability policies of the UK come squarely from the European Union. In addition, DG Sanco and DG Environment of the European Commission, the two key DGs for GMOs since the end of the 1990s, became instrumental for the environmental protection network. As explained above, the farm-scale evaluations anticipated the requirements of Directive 2001/18. The adoption of this directive by the European Union was also promoted by the actors of the environmental protection network and Michael Meacher, the environment minister (Levidow and Carr 2000: 4). Without the European Union's capacity to influence GMO policy, it is far from certain that Meacher's cautious perspective would have prevailed over Blair's optimism. As this example shows, the European Union can increase the importance of some domestic networks of actors, who have preferences distinct from the prime minister, and who are otherwise without influence (Montpetit 2000). In short, the European Union, in addition to the creation of autonomous administrative agencies, worked to undermine the Whitehall model.

## CONCLUSION

This chapter demonstrates that policy networks matter a great deal in the explanation of policy designs. In fact, the UK's GMO and ART policies differ sharply, as do policy networks in both sectors, and this is no mere coincidence. The permissiveness of the UK's ART policy is largely explained by the incentives the Powell bill created in 1985 among the medical and scientific communities, the key target groups of the bill, to join forces in a cohesive policy network with tight interconnections to decisive state actors. In contrast, opponents lost much of their cohesion and interconnections with decisive state actors during the 1980s, forming a weak issue network in the 1990s. Given its weakness, this issue network failed to convince British policy-makers to adopt an understanding of ART, distinctive from that of target groups, that would have granted human embryos the same rights as human beings. These network conditions were conducive to the adoption of a permissive ART policy.

In the GMO sector, the network of technology-concerned actors has been more successful at changing policy-makers' understanding of the issue. Enjoying increasing cohesion in the context of regulatory failures in scientific sectors (notably mad cow disease), technology-concerned actors were successful at convincing policy-makers that they needed to broaden their understanding of risk assessment. Confined to agronomic risks in the early 1990s, the understanding of risk assessment that prevailed at the end of the 1990s also embraced ecological risks, thanks to a tight interconnection the network provided to a sympathetic environment minister and department. In contrast, the cohesion of the pro-technology network, the main target of GMO policy, began to erode, as some industries feared profit losses when consumers began to adopt the discourse of technology-concerned actors. These network conditions were conducive to the adoption of a restrictive GMO policy.

Networks evolve in an institutional context, which proved influential in the adoption of a restrictive GMO policy design and a permissive ART policy design. The success of the policy community of target groups in the ART sector was partly attributable to the concentration of power afforded by the British Whitehall system of government during the 1980s. Granting the prime minister a great deal of power over her party and policy-making more generally, the Whitehall model prevented the adoption of the Powell bill in 1985 and bought precious time to the policy community of targets to develop and present arguments justifying the adoption of a permissive bill. In contrast, when the network of technology-concerned actors became cohesive in the mid-1980s, changes in the Whitehall model had been undertaken, notably through the creation of autonomous agencies. Even though the Labour prime minister first elected in 1997 was a biotechnology enthusiast, he did not have the power Thatcher had in the 1980s to help the community of target groups translate its preferences into policy. The European Union further contributed to the erosion of the power of the prime minister and helped technology-concerned actors obtain a restrictive GMO policy.

## REFERENCES

Bernauer, T., and E. Meins. 2003. Technological Revolution Meets Policy and the Market: Explaining Cross-national Differences in Agricultural Biotechnology Regulation. *European Journal of Political Research* 42 (5): 643–83.

Biotechnology & GMOs Information Website. http://gmoinfo.jrc.it.

Blank, R. H. 2004. The United Kingdom: Regulation Through a National Licensing Authority. In *Comparative Biomedical Policy: Governing Assisted Reproductive Technology.* Eds. I. Bleiklie, M. L. Goggin, and C. Rothmayr. Pp. 120–37. London: Routledge.

Festing, S., J. Gillott, and J. Tizzard. 2003. *A Guide to Organisations Critical of Human Genetics and Embryology.* London: Association of Medical Research Charities, Genetic Interest Group and Progress Education Trust.

Gunning, J., and V. English. 1993. *Human In Vitro Fertilisation: A Case Study in the Regulation of Medical Innovation.* Darmouth: Aldershot.

Levidow, L., and S. Carr. 2000. UK: Precautionary Commercialization? *Journal of Risk Research* 33 (3): 187–285.

Marsh, D., D. Richards, and M. J. Smith. 2001. *Changing Patterns of Governance in the United Kingdom: Reinventing Whitehall?* Houndmills: Palgrave.

———. 2001. Understanding Policy Networks: Towards a Dialectical Approach. *Political Studies.* 41 (1): 528–41.

Marsh, D., and M. Smith. 2000. Understanding Policy Networks: Towards a Dialectical Approach. *Political Studies* 48 (1): 4-21.

———. 2004. Policy Networks, Federalism and Managerial Ideas: How ART Non-Decision in Canada Safeguards the Autonomy of the Medical Profession. In *Comparative Biomedical Policy: Governing Assisted Reproductive Technology.* Eds. I. Bleiklie, M. L. Goggin, and C. Rothmayr. Pp. 64–81. London: Routledge.

Montpetit, É. 2000. Europeanization and Domestic Politics: Europe and the Development of a French Environmental Policy for the Agricultural Sector. *Journal of European Public Policy* 7 (4): 576–92.

Montpetit, É., C. Rothmayr, and F. Varone. 2005. Institutional Vulnerability to Social Construction: Federalism, Target Populations, and Policy Designs for Assisted Reproductive Technology. *Comparative Political Studies* 38 (2): 119–42.

Mulkay, M. 1995a. Galileo and the Embryos: Religion and Science in Parliamentary Debate over Research on Human Embryos. *Social Studies of Science* 25 (3): 499–532.

———. 1995b. Political Parties, Parliamentary Lobbies and Embryo Research. *Public Understanding of Science* 4 (1): 31–55.

Nuffield Council on Bioethics. 1999. *Genetically Modified Crops: The Ethical and Social Issues*. London: Nuffield Council on Bioethics.

Pollitt, C., and G. Bouckaert, 2000. *Public Management Reform: A Comparative Analysis*. Oxford: Oxford University Press.

Toke, D., and D. Marsh. 2003. Policy Networks and the GM Crops Issue: Assessing the Utility of a Dialectical Model of Policy Networks. *Public Administration* 81 (2): 229–51.

UK Agricultural and Environment Biotechnology Commission. 2001. *Crops on Trial*. London: Agricultural and Environment Biotechnology Commission.

UK Cabinet Office. Office of Science and Technology. 1999. *The Advisory and Regulatory Framework for Biotechnology: Report from the Government's Review*. London: Cabinet Office.

UK Department for Environment, Food and Rural Affairs. Farmscale Evaluations Research Team and the Scientific Steering Committee. 2003. *GM Crops: Effects on Farmland Wildlife*. London: Department for Environment, Food and Rural Affairs.

UK Food Standards Agency. 2003. *The Food Standards Agency's Contribution to the Public Dialogue: Consumer Views of GM Food*. London: Food Standards Agency.

Warnock, M. 1984. *Report of the Committee of Inquiry into Human Fertilisation and Embryology*. London: Department of Health & Social Security, Her Majesty's Stationery Office.

## 6

# POLICY MEDIATION OF TENSIONS REGARDING BIOTECHNOLOGY IN FRANCE

*Éric Montpetit (Université de Montréal) and*
*Nathalie Schiffino (Université catholique de Louvain)*

The French economy has been one of the strongest in the postwar period due in part to the county's industrial policy, which has historically encouraged technological progress (Hall 1986). The governing elite, often trained in the country's top engineering schools, are naturally inclined to be favorable towards policies that promote most kinds of scientific advances. Biotechnology has, however, been a source of controversy and has created tension among the French citizenry. Biotechnological progress promises great economic potential, but also carries significant risks that have become worrisome to a significant proportion of French citizens. Biotechnology professes to transform agriculture and food, but many citizens wish to see these cherished sectors protected from modern progress. Given the strong disagreements over biotechnological issues between French citizens, as well as between citizens and the elite, the development of biotechnology policies in France is particularly challenging.

This chapter presents an in-depth analysis of French policy development for assisted reproductive technology (ART) and genetically modified organisms (GMO), two biotechnologies situated on opposite sides of the French biotechnology debate with ART receiving far more support than GMOs. Despite this difference, policy-makers have designed policies that do not sharply diverge. We argue that differences in policy networks, combined with key characteristics of the French polity between 1995 and 2004, provide powerful explanations for the similarities and differences in the country's ART and GMO policies. Although present, international forces did not exert as much influence on policy-making as these domestic factors.

## INTERMEDIATE ART POLICY

The French regulation consists of a mix of permissive and restrictive measures.[1] Although access to fertility treatments are encouraged by being covered by the country's health insurance system, notably of IVF up to four cycles, the law strictly limits access to fertility clinics to stable heterosexual couples. This mixture of restrictive and permissive measures is the result of a policy-making process that began in the 1970s after the creation of the first sperm and egg bank (Centre d'étude et de conservation des oeufs et du sperme, CECOS) and that accelerated in the 1980s after the birth of the first French IVF baby. In light of these developments, physicians and researchers created local ethics committees to provide advice on the ethics of biomedical research projects and to develop the protocols for therapeutic trials. Although nonphysicians could serve on them, these committees represented a self-regulatory apparatus insofar as they mostly involved medical professionals, despite the fact that the Huriet legislation of 1988 changed the name of these committees to "comités consultatifs de protection des personnes qui se prêtent à des recherches biomédicales" in order to redefine them explicitly as having a role in protecting persons who agree to participate in biomedical research. The first intervention by the national government, namely the creation of the "Comité consultatif national d'éthique" (CCNE) by a 1983 presidential decree, did not significantly alter this self-regulatory approach. In fact, the committee, while mandated to examine the ethics of ART, was formed mostly of medical specialists and its first president, Jean Bernard, was a physician himself. In the 1980s, the "décrets Barzach" enabled the creation of the ad hoc "Commission nationale de médecine et de biologie de la reproduction," which became in 1994 the "Commission nationale de médecine et de biologie de la reproduction et du diagnostic prenatal" (CNMBRDP). The results of these decrees were also largely consistent with self-regulation. Focusing on the sanitary and medical specialty requirements for operating fertility clinics, these decrees never went as far as suggesting which treatments were admissible and which were not. This situation greatly encouraged the development of fertility clinics and normalized the use of fertility treatments across France. In 1998, 18 percent of childbirths to couples involved a fertility treatment and the proportion continues to rise.

Within this ART-friendly climate, the government adopted three bioethics laws in 1994. These laws aimed at imposing legislated limits on what can be achieved in fertility clinics and in laboratories. Act 94-654 explicitly defines ART to include embryo transfers or any technology aimed at procreation. Unlike earlier ART policies, this act interferes with the collective autonomy of the medical profession by determining under which conditions fertility treatments can be performed. Rather restrictive in comparative terms, the measures contained in the act seek to prevent any commercialization of gametes and embryos. The act thus imposes strict conditions on gamete donations and on the use of donations: donations have to be free; donors have to belong to a couple with children; and donors must remain anonymous. The beneficiaries of the donations should be in stable heterosexual couples, and their use of donated gametes should be "absolutely necessary." The act also imposes strict conditions for the practice of prenatal diagnosis, which is limited to embryos from couples whose heredity increases the risks of passing a genetic disease to their children. If some of these measures are restrictive, it should however be underlined that they are not aimed at discouraging couples from resorting to ART. To the contrary, the act portrays ART as a legitimate, even promising, scientific solution to fertility problems. For example, the act never calls into question the coverage of treatments in the national heath insurance system, but rather encourages greater access to infertile couples. By failing to provide any framework to assess and authorize treatment experiments and trials on humans, legislators revealed a strong faith in science. Thus, Intra-Cytoplasmic Sperm Injection (ICSI), a treatment developed in Belgium in 1991, appeared in French fertility clinics in 1995 without any public scrutiny of its risks and benefits.

If the bioethics laws legitimized and encouraged fertility clinics, the laws contained strict prohibitions with regard to embryo research. When the acts were developed in 1994, embryo research was essentially viewed as a means to improve fertility treatments. The legislators decided, out of respect for the sacred character of embryos, that it was unethical to use embryos in research for the purpose of allowing improvements in the maturation success of future embryos, since embryos used in research are destroyed. In other words, French legislators decided that the principle which suggests that it is unethical to sacrifice a human life to improve that of another human should also apply to embryos. This principle, in addition to the fear of eugenics, justified forbidding the creation of embryos solely for research purposes, as well as any research modifying their genetic potential. This restriction, in fact, eliminates nearly any possibilities of conducting research on embryos. In 2000, the CNMBRDP, whose responsibilities were extended to approving embryo research projects under the bioethics laws of 1994, received only eleven applications. Of these eleven applications, only five were approved (Charles and Claeys 2001: 117).

Embedded in the 1994 bioethics laws is the obligation to revise its measures after five years of application. This revision was therefore undertaken by the Jospin government in 1999 and largely focused on the restrictions on embryo research. By the end of the 1990s, embryos had become an important source of stem cells, thus encouraging a separation between the issues of fertility and embryo research. Although embryo research resorts to ART insofar as the creation of embryos are concerned, it does not directly pertain to fertility but rather relates to genomics and the treatment of incurable diseases. Embryo research therefore raises the new question of whether it is ethical to use embryos to improve the lives of living human beings. The socialist majority in the "Assemblée nationale" appeared inclined to encourage scientists to pursue such research, albeit under controlled conditions (Charles and Claeys 2001: 133). However, the defeat of the Socialist Party in the 2002 presidential and legislative elections halted the attempt to render the legislative measures pertaining to embryo research more permissive.

Ten years after the adoption of the first bioethics laws in 1994, the lengthy revision process ended with the 2004 law. Unlike what the Socialist government had planned, the 2004 law did not increase the permissiveness of French ART policy design. First, the law prohibited reproductive and therapeutic cloning and imposed severe sanctions on any violator, going so far as to treat reproductive cloning as a crime against humanity. Second, embryonic research was, in principle, prohibited. The law, however, authorized some embryonic research during a five-year period. To carry out embryonic research, researchers now have to demonstrate that the purpose they wish to pursue cannot be attained by other means; they have to demonstrate that their research will produce "major" therapeutic progress; and they have to carry out their research on leftover embryos. Third, the law created the "Agence de biomédecine," whose responsibilities included approving research projects, licensing clinics and approving their practices. The agency's mission was similar to that of the UK's Human Fertility and Embryology Authority, but applied in the context of a far more restrictive act when it comes to embryonic research. The permissiveness of the law becomes more apparent when one moves from laboratories to fertility clinics. In terms of the conditions of fertility practices, excluding those related to gamete donation and prenatal diagnosis, the law imposed few constraints

**Table 6.1. Current French ART Policy Design**

| | Research | Practice of fertility treatments | Measures aimed at patients |
|---|---|---|---|
| Policy Design | - Creation of embryos prohibited<br>- Embryo research generally prohibited, although with some exceptions<br>- Therapeutic cloning strictly prohibited<br>- Approval by the Biomedicine Agency required | - License by the Biomedicine Agency for physicians and clinics required<br>- Gamete donations have to be free, anonymous, and from couples with children<br>- Prenatal diagnosis permitted only for couples at risk of transmitting a serious genetic disease | - Access limited to stable heterosexual couples diagnosed with an infertility problem<br>- Full funding provided under the national health insurance system |
| Type | Restrictive | Intermediate: permissive for medical practices except for prenatal diagnosis; restrictive for gamete donations | Intermediate: some sociological restriction for access; permissive funding |

on the autonomy of medical professionals. Table 6.1 summarizes the French ART policy after the 2004 revision of the bioethics laws, which resulted in an ART policy design that is moderate overall, despite imposing tough restrictions on embryo research.

## INTERMEDIATE GMO POLICY

The GMO policy design, originally permissive, has, just like the ART policy design, come to include some significant restrictions. In contrast to the current ART policy design, however, the current design for GMOs contains stricter restrictions on use and consumer protection than on research.

Research projects were at the origin of the policy process for GMOs. In 1974 a coalition of biochemists and molecular biologists called for a moratorium on recombinant DNA research until risks for researchers and laboratory staffs were better known. This French controversy was quickly transposed internationally at the 1975 Asilomar Conference (United States) where scientists asked for the establishment of adequate risk management arrangements. Following this conference, French scientists, supported by the Ministry for Research, took it upon themselves to create such an arrangement—the "Commission de classement et de contrôle des expériences du génie génétique." As a mechanism of collective self-regulation by scientists, the commission was given the responsibility for risk assessment and risk management for the contained use of GMOs. These responsibilities were transferred to the "Commission du génie génétique" (CGG) by a 1989 government decree, which however did not alter the operations of risk assessment and management for the contained use of GMOs.

The OECD had begun in 1983 to devise risk assessment and risk management arrangements for the environmental release of GMOs, leading to the publication of a "Blue book" in 1986, which established the principle of gradual release. In France, the mandate to implement this principle was given to the ad hoc "Commission du génie biomoléculaire" (CGB) also created in 1986 by ministerial decision. The creation of a new institution for regulating GMOs, it should be noted, corresponds to a process-based approach whereby genetic modification is in itself considered sufficiently risky an activity to deserve its own regulatory apparatus. This approach contrasts with the product-based approach, described in chapters 3 and 4, whereby the products of genetic modification are regulated by the relevant institutions. Until 1992, CGB authorizations for field trials of GMOs were voluntary, but all firms did, in fact, submit their projects to the organization (Chevassus-au-Louis 2001: 48). It should also be underlined that the CGB, unlike its counterpart for contained used, includes some nonscientists: a judicial expert, a representative for the environmental movement, a representative for consumers, a representative for industry and a member of Parliament. Nevertheless, studies have shown that scientists not only outnumbered other participants, but they entirely dominated the discussions, imposing their scientific language and thereby silencing lay representatives during the evaluation of any GMO (Roy 2001). Abiding by such tight scientific criteria, the CGB has authorized more field trials for GMOs than any of its counterparts in the rest of Europe (Joly et al. 2000: 28). In order to translate the European directives 90/219/CEE and 90/220/CEE into national law, the French government passed two acts in 1992 (92-654 and 76-663). These two acts legally recognized the CGB and the CGG, thereby formalizing the prevailing risk assessment and management practices for both contained use and environmental release of GMOs.

The general climate of permissiveness toward GMOs in France began to change in 1996 after a series of events that had been widely reported in the media. First, a number of well-known scientists, physicians and other health professionals signed a petition in 1996 demanding a moratorium on the commercialization of food-related GMOs until tougher regulations were put in place. Second, American ships filled with soybeans, some of which had been grown from GM seeds, began to arrive in European ports in 1996. Greenpeace organized a number of operations to prevent the unloading of these ships. Covering these operations, *Libération*, a well-known newspaper, published a front-page article entitled *Mad Soy Alert* and the article had an effect on a public concerned about mad cow disease. A number of food safety scandals from across Europe, in particular mad cow disease, had by this point created a shock wave in French public opinion. To reassure the public, the French government passed a law in 1998 that ensured a close monitoring of food safety. The law created the "Agence française de sécurité sanitaire des aliments" (AFSSA), which received the responsibility for assessing the health impact of genetically modified organisms likely to enter the food chain, thereby confirming the possibility of a connection between food safety and GMOs. And, in a third case, governmental hesitations over the authorization of a Bt corn confirmed increased uneasiness in France regarding GMOs. After having introduced Novartis's Bt corn into the European approval procedure defined by 90/220/CEE, the Juppé government authorized its commercialization on February 5, 1997. Seven days later, however, the same government decided to prohibit the cultivation of the same corn seeds for environmental reasons. On the 13th of the same month came the resignation of the president of the CGB, whose advice on GMOs, up until then, had always been followed. In legislative elections a few months later, Jospin's socialists defeated the Juppé right-wing government. With the controversy around the Bt corn decision still alive, the new government chose in November 1997 to reverse its predecessor's prohibition, but simultaneously announced the organization of a citizen's conference to inform government policy on GMOs. Shortly thereafter, the inscription of the Bt corn on the country's seed registry prompted Greenpeace to launch a judicial challenge against the government. Against all odds, the "Conseil d'État" accepted Greenpeace's argument, thereby canceling the registration of the Bt corn on the seed registry. In December of the same year, however, the European Court of Justice finally stated that the "Conseil d'État" did not have the jurisdiction to make such a decision, thereby reaffirming the legality of Jospin's decision to allow the cultivation of Bt corn in France. In response to these hesitations, a number of policy decisions were produced in order to redress a shaken public confidence in the adequacy of the French regulatory process for GMOs.

As mentioned above, the Jospin government decided to mandate the "Office parlementaire des choix scientifiques et technologiques"

(OPECST) to organize a citizen's conference. The OPECST chose to follow the Danish consensus conference model whereby a panel of citizens is randomly selected to be exposed to conflicting opinions before deliberating on a particular policy issue. In the French case, the panel was due to report to the president of the OPECST, Jean-Yves Le Déaut, who had been given the responsibility to draw conclusions for the government. While adopting a moderate position toward GMOs, the citizen's panel, and subsequently Le Déaut, insisted on the importance of close monitoring and transparency. This advice entailed a reform of existing regulatory mechanisms, notably to enable the full expression of scientific controversies and the development of a traceability policy. Thus, in July 1998, a ministerial decision revised the CGB to make sure that the organization reflected scientific controversies. Appointments to the CGB were thus made to integrate scientists who had critical views on GMOs. In addition, the functioning of the organization was revised to allow for debates and the diffusion of minority opinions. Although the recommendations of the CGB needed to be carefully justified, the minutes of the organization's meetings also needed to clearly present minority opinions (Joly et al. 2000: 45; Roy 2001). In November 1998, the "Comité de biovigilance" (CB) was created to closely monitor any approved GMO. Even more than the CGB, the CB's composition should encourage the expression of scientific controversies (Joly et al. 2000: 44). Lastly, the French environment minister played a key role at a June 1999 European Council meeting of environmental ministers in forming a coalition of five member states demanding a revision to labeling regulations and the devel-

**Table 6.2.   Current French GMO Policy Design**

|  | Field trials | Commercialization and cultivation | Consumer and environmental protections |
|---|---|---|---|
| Policy Design | - Government approval required, following the advice of the reformed CGB and the "Comité de Biovigilance" | - De facto moratorium on the marketing of new GMOs (1999–2004)<br>- Government approval required, following the advice of the reformed CGB and the AFSSA | - Labeling requirements defined at the European level<br>- Traceability requirements defined at the European level |
| Type | Intermediate | Restrictive | Restrictive |

opment of traceability rules before approving any additional GMO (Rules 1829/2003 and 1830/2003). This coalition was sufficiently large to cause a de facto moratorium on the commercialization of GMOs (1999–2004).

France remains extremely late in transposing the European directive 2001/18 on field trials and commercialization. The European Court of Justice has ruled against France for not complying with its duties. A law is currently in preparation in Parliament; however, faced with a reactive public opinion, political decision-makers seem reluctant to proceed with GMO legislation. This situation effectively prevents commercialization, even if authorizations for field trials can be given.

These policy decisions have increased the restrictiveness of the French GMO policy, although very few changes occurred at the level of contained research. While the reformed CGB still approves field trials, the organization's composition and functioning rules increased the difficulty for biotechnology firms to obtain such approvals. The most significant restrictions apply to the commercialization and cultivation of GMOs, which are currently under what French authorities say is a temporary prohibition that will be lifted once the European Union puts in place acceptable labeling and traceability requirements (such regulations were finally adopted in 2005). Lastly, the GMOs approved for marketing before the moratorium are under the close scrutiny of the CB, which is not afraid of initiating controversy.

## EXPLAINING THE FRENCH ART AND GMO POLICY DESIGNS

The above analysis suggests a mix of commonalities and differences between ART and GMO policy designs. On the one hand, both designs became gradually more restrictive. On the other hand, both designs differ in that restrictions apply to different aspects in each sector. While restrictions became stricter for research pertaining to ART, French policies remained relatively permissive for GMO research, notably field trials. Conversely, restrictions applying to the use or the commercialization of the results of these researches are stricter for GMOs than they are for ART (in the sense that health coverage is broad). In what follows, we provide an explanation for these commonalities and differences. We argue that similarities and dissimilarities in policy networks partly account for divergence and convergence. In the ART sector, a variety of actors participated in the development of the ethics laws and no cohesive policy community could effectively prevent the adoption of restrictive measures, which had been preferred by the Conservative governments of both 1994 and 2004. In fact, because these Conservative governments were not beholden to coalition partners and controlled both the presidency and the National Assembly, their concern for the ethics of embryo research were easily translated into a restrictive policy. These Conservative governments also promoted family values and therefore were favorable to fertility treatments for heterosexual couples, hence the overall moderate ART policy design. In contrast, Jospin's Socialist government, which made significant GMO policy decisions, was trapped between its coalition partner, the Green Party, and a powerful policy community of target groups. The government thus favored a moderate policy regarding field trials. Restrictions concerning the commercialization of GMOs were, as in the UK, closely associated with the European Union.

## Policy Networks

Policy networks are related to policy outcomes in various ways. In this book, we have suggested that permissive policies can be related to closed and cohesive policy communities of pro-technology actors. We further argue that networks interconnecting decisive state institutions and civil society actors concerned about technological advances can enable the adoption of restrictive measures. It is also useful, we show in this chapter, to treat policy networks as a structure, constructed over time, which cannot be manipulated at will and whose evolution is not always in tune with the social, economic and political context. It has therefore been argued that networks have a mediating effect between civil society and the state (Coleman 1998). If the social and political context can change rapidly, some networks can have a moderating effect on the pace and depth of policy change. The French situation, in the area of GMO in particular, illustrates this dynamic well.

If the mediating effect of policy networks is particularly significant in the area of GMOs, it is because public views on the matter have hardened significantly since the mid-1990s. In 1996, 54 percent of French citizens were favorable to GM food. In 2002, the figure was 30 percent. For GM crops, the decline in public support is also significant, falling from 79 percent in 1996 to 55 percent in 2002. In comparison, support for genetic testing fell by only four percent, varying between 96 percent in 1996 and 92 percent in 2002 (Eurobarometer 58.0).

Prior to the mid-1990s, GMO policy was largely the purview of a policy network closed around a limited number of actors sharing cohesive ideas. At the center of the network was of course the CGB, at the time dominated by scientists who shared a homogeneous perspective anchored in molecular biology (Roy 2001). These scientists had frequent contacts with the researchers of the "Institut national de recherche agronomique" (INRA), as well as with firms involved in the development of GMOs. The CGB was also closely interconnected with the agriculture ministry. The actors in this network generally embraced a paradigm of scientific progress that was mostly favorable to biotechnology, even if the CGB occasionally recommended against the approval of a particular GMO to build its credibility in the public. As Joly and his colleagues (2000: 29) wrote:

> Created on a philosophy of scientific progress, the objective of the CGB is to promote the use of GMOs. To do this, this institution needs a credible regulatory framework. But the framework does not suffice. The CGB must also approve plants that benefit consumers to stimulate the public acceptability of these new plants. Thus, in 1994, the French authority blocked the approval of herbicide tolerant colza.

In short, between 1986 and the mid-1990s, a cohesive policy community, providing a tight interconnection to the state through the CGB and the agriculture ministry, dominated France's relatively permissive GMO policy.

The first event that stimulated the GMO controversy that took off in the mid-1990s was the increasingly visible divisions within the scientific community on the potential risks and the expected benefits of GMOs. Internal disputes in prominent places such as the INRA were made public, notably following the "Appel des scientifiques, des médecins et des professionnels de la santé pour un contrôle des applications du génie génétique" (May 1996). The relatively quiet functioning of the policy community of pro-technology actors was deeply disrupted when the Juppé government decided to approve the commercialization of Novartis's Bt corn, but shortly thereafter prohibited its cultivation, against the advice of the CGB (February 1997). Up until then, the CGB had worked closely with the agriculture ministry and Matignon, the prime minister's office, had stayed outside of this area. In any case, Matignon's decision encouraged civil society actors, such as the "Confédération paysanne" and Greenpeace, in their mobilization efforts against the scientific paradigm that was informing the network's perspective. In addition, the decision prompted the involvement of new state actors, with fresh perspectives on the issue: the "Conseil d'État" and the OPECST through a citizen conference (Joly et al. 2000: 48). Changes in the French policy design became unavoidable.

As we saw, the revision of the CGB's composition and the creation of a "Comité de biovigilance" were among the key changes. Moreover, a sanitary agency, AFSSA, was created. As indicated in table 6.2, however, this change only made the French policy regarding field trials of GMOs slightly more restrictive. It moved from a permissive to an intermediate nature. Although the CGB currently includes members with critical perspectives of GMOs, pro-technology actors are still in a majority position in the institution. The Comité de biovigiliance may very well be a more critical institution, but its policy responsibilities are far less significant than those of the CGB. In other words, the policy change did not perfectly reflect the change in the attitudes of French citizens toward GMOs. The cohesive pre-1995 pro-technology policy network mediated the policy change demanded by citizens via technology-concerned actors. The French government simply could not do away, in one sweep, with a network of actors who had exercised, for a period of over ten years, key responsibilities over the formulation and the implementation of the country's GMO policy.

French public opinion did not radicalize against ART as it did over GMOs. The coverage of infertility treatments under the country's health insurance system and most forms of research involving embryos were never vehemently opposed by civil society. In the years leading to the adoption of the 1994 and 2004 bioethics laws, most citizens trusted the networks of scientists and physicians making decisions over ART. In this area, however, policy networks were, from the beginning, not as cohesive as the pre-1995 GMO policy community.

As mentioned above, one of the first decisions of the French government in the ART sector was the creation of the CCNE in 1983. The CCNE was an advisory body mandated to think about bioethics. Therefore, it differs significantly from the CGB, which was given regulatory responsibilities. Regulatory responsibilities and the immediate significance of CGB's decisions attracted actors with key stakes in GMOs and encouraged the institution to associate itself with actors sharing its views, if only to make decisions effectively and as legitimately as possible. Given the nature of its mission, the CCNE did not play the same kind of role in the creation of a cohesive network. The creation of the "Agence de biomédecine," however, may very well act on networks in the same manner as the CGB in the GMO sector although it is too early to tell. Meanwhile, in the area of ART in France, networks are closer to issue networks. An interviewee illustrates well the network situation prevailing at the end of the 1990s when the government was revising the 1994 bioethics laws:

> We had to think, for example, about stem cell research. Stem cell research per se concerns only a very small number of laboratories, mostly related to the Institut national de la recherche médicale or to the Institut national de la

recherche scientifique. All this represents a very small number of units and persons, already facing restrictive regulations. The matter could have been resolved easily among these people. However, ART is not only about laboratory research; laboratory research sometimes transfers into hospitals. At this scale, it involves dozens of institutions, often more knowledgeable in appendicitis surgery than in ART. The perspectives at this scale differ because they are informed by different practices and do not apply to the same objects. When you are in a laboratory, your interest is in the short term; when you are in a hospital you want to think about the long-term implications of laboratory discoveries for your practice.[2]

Organized groups were also involved in the work leading to the revision of the 1994 bioethics laws. Patients' associations, feminists' groups and organizations representing the disabled were invited to make presentations during parliamentary hearings. These groups often sided with those practitioners who argued in favor of a cautious approach and whose perspective often differed from that of laboratory scientists. In a manner that resembles the UK case, several French ART clinicians chose to cooperate with policymakers and accepted some restrictions on their practices (Montpetit, Rothmayr, and Varone 2005). In should be underlined, however, that the Conservative governments that adopted the first and then the revised bioethics laws were naturally inclined towards policies favorable to traditional family values. Consequently, they were favorable to fertility clinics, as long as they provided services to infertile heterosexual couples.

In the next section, we contend that the Conservative governments of 1993 and of 2004 were, institutionally speaking, in a strong position to translate their preferences regarding embryo research and fertility treatments in the first and revised bioethics laws. The network situation contributed to the two governments' capacity to enact restrictions on embryo research, insofar as laboratory scientists were not interconnected with decisive state institutions to the extent that pro-technology actors were in the GMO sector. Nor did the medical community form a cohesive network. Networks in the ART sector were issue networks. Had the government faced a network situation comparable to that prevailing in the area of GMOs, restrictions on embryo research would have been much more difficult to adopt.

Table 6.3 summarizes the network situations in both sectors. In the next section, we contrast the institutional situations prevailing during the development of the bioethics laws with that prevailing during the period of GMO policy redesign. These institutional situations, along with policy networks,

Table 6.3. Contrasting Network Situations between ART and GMO

| | Pro-technology networks | | Networks of actors concerned about technology | |
|---|---|---|---|---|
| | Medical network (ART) | Industry network (GMO) | Patient/Feminist networks (ART) | Environmental/Consumer network (GMO) |
| Cohesion | Low | High | Moderate | Moderate |
| Interconnectedness | Moderate | Tight | Loose | Moderate |

had a significant policy impact. Table 6.3, however, already points to an explanation as to why the policy regarding GMO field trials never became as restrictive as the policies for embryo research.

## Country Patterns

The semi-presidential system of France enables a single party to, on occasion, entirely dominate policy-making; at other times it requires a coalition government; and yet at other times it possesses some of the attributes of the United States' system of separation of powers. As in the United States, French electors can vote separately for legislators and the president; they need not be from the same party, but the parallel stops here. Indeed, when elections yield a divided government in France, the responsibility to form the government is no longer exercised by the president. Such an institutional situation is known as cohabitation. In such a case, the leader of the party that won the largest number of candidates in the legislative assembly is appointed prime minister and in situations of divided government he or she appoints the government, although the president still chairs cabinet meetings. When the prime minister's party does not have a clear majority of seats in Parliament, he or she may have to find one or several coalition partners among the minority parties and appoint cabinet ministers from this or these parties. In other words, the institutional context in France varies considerably over time, depending on electoral outcomes. At times, the institutional context is one of power concentration in the hands of the president, but at different times it provides veto points to other actors within the executive branch. As we argue below, the institutional context in the ART sector was one of power concentration, facilitating the translation of the policy preferences of Conservative governments into policy. In the GMO sector, the Socialist government of Prime Minister Jospin, in addition to cohabiting with President Chirac, had to deal with a coalition partner with strong environmental preferences.

In the spring of 2002, Jacques Chirac was reelected as president of the Republic. A few weeks later, his conservative political party, the "Union pour un mouvement populaire" (UMP), won a majority of seats in the "National Assembly," earning Chirac the possibility to appoint a government. Under these circumstances, Chirac and his UMP entourage had significant control over policy-making. The 2002 elections marked the end of five years of cohabitation, during which Chirac was forced, because his party did not control a majority of seats in the "National Assembly," to abandon several of his policy responsibilities to Jospin's Socialist government. Jospin's Socialists themselves did not have sufficient seats in the legislative assembly to govern alone and therefore had to allow the formation of a coalition government with the Green Party. In other words, over the last fifteen years, French political parties were variously capable of translating their preferences in public policy.

In the area of ART, the crucial years were those leading to the adoption of the bioethics laws in 1994. The first bills on ART were introduced in the "National Assembly" in 1992 by a Socialist government well positioned to obtain swift passage of its legislation. Designed by a Socialist government and a presidency comprising several bio-optimists, these bills were rather permissive, legitimizing several aspects of ART research. However, the Socialist Party was not unified behind this permissive preference and therefore the bills stirred significant legislative debates. Following these debates and public consultations, the ad hoc commission mandated by the "National Assembly" to study the bills proposed some restrictive amendments. The decisive move toward restrictions on research, however, came in 1993 when legislative elections brought a right-wing majority to the assembly. Unlike members of the Socialist Party, supporters of the Right in France are more unified in the view that embryos deserve protection; they oppose any change in the current sacred status accorded to embryos by the country's abortion law (Engeli 2004). After the 1993 elections, parliamentary debates shifted towards the importance of protecting embryos in a manner consistent with the 1975 law on the interruption of pregnancy. This perspective justified imposing restrictions on research leading to the destruction or genetic alteration of embryos, an essential element of the 1994 laws on bioethics.

As mentioned above, the laws included the obligation to revise itself after five years. When the five-year period arrived in 1999, Jospin's Socialists, in coalition with the Green Party, were in power. A similar sequence of events repeated itself. Comprising several bio-optimists, the government sought to increase the permissiveness of the 1994 laws on bioethics, notably its measures on embryo research. The proposed amendments launched a long and cumbersome debate in the National Assembly as well as within the executive branch. Matignon, the prime minister's office, initially held very permissive preferences on embryo research, even accepting therapeutic cloning, while Chirac, a Conservative president, vehemently opposed all forms of cloning. Even if Matignon became increasingly open to a compromise, the adoption of the bill before the 2002 elections was impossible. After losing the election, the Socialist Party was forced to abandon control over the revisions of the laws on bioethics to a strong UMP, a conservative party. Ironically, Jean-François Mattei, who played a key role in the adoption of restrictive measures on ART research in 1994, was appointed as the health minister and thus became responsible for the laws on bioethics. In the end, the review of the 1994 bioethics laws, carried out under an undivided right-wing government, included the adoption of restrictive measures.

If the influence of political parties on making ART research policy is undeniable, their role over GMO is far less obvious. As we just saw, ART policy-making activities were marked by a transition between a socialist and a conservative government. Conversely, most GMO policy-making activities occurred during a transition between a strong conservative government and a period of cohabitation between Chirac's presidency and Jospin's red-green coalition government. As we saw above, the February 1997 contradictory decisions made by Alain Juppé, Chirac's prime minister, about Novartis's Bt corn fed into an important controversy. This controversy was a few months later relayed to the newly elected government of Jospin, who had preferred a relatively permissive biotechnology policy under normal conditions, but who had to cope with cohabitation and a coalition partner. Jospin thus decided to avoid engaging his government into a regular legislative or governmental process to deal with the GMO issue, instead announcing a citizen's conference. Although organized by Parliament's OPECST, the conference effectively removed the issue from the hands of political parties. This decision, as we saw above, contributed to a slight opening of what had been, prior to the mid-1990s, a cohesive policy community.

In short, the Conservative or Right-wing party contributes to the explanation of the French policy design in a single sector: ART. Although, as we saw above, networks provide a coherent explanation of why governments were more able to make policy in the ART sector, divisions within the Socialist Party over ART harmed its capacity to render the policy more permissive when it had the opportunity to do so. Furthermore, the Right was able to adopt restrictions in the ART sector under very specific institutional conditions. Indeed, it did so outside cohabitation and when it did not need a coalition partner. In other words, political parties matter in France under very specific institutional conditions. Indeed, the socialist government was able to remove the GMO issue from the control of political parties, seeking the advice of OPECST and a citizen's conference.

## Internationalization

Biotechnological innovations are often anchored in transnational networks of firms and research centers. When biotechnological innovations occur within individual firms, whose activities are in a single country, they are nevertheless quickly diffused internationally. The cloning of an animal first occurred in Scotland; shortly thereafter, however, researchers in several other countries began to clone animals. The transnational character of biotechnology has motivated policy-makers to participate in, or at least pay attention to, international governance arrangements in this sector. Such a system of governance, it has been argued in the public policy literature, multiplies the opportunities for communicative action, thereby encouraging the construction of transnational norms policy transfers and policy learning at the domestic level (Risse-Kappen 1996; Héritier 1999; Kohler-Koch 1996; Radaelli: 2000). We now turn to the question of the extent to which ART and GMO policy designs in France can be associated with such internationlization processes. We argue that the moderate increases in the restrictiveness of biotechnology policy discussed above come neither principally from an integration of international norms, nor from the emulation of other countries' policies. However, actors concerned about GMOs have successfully used the European Union to obtain restrictions on commercialization.

In the area of ART, several international institutions can be relevant: the World Health Organization, UNESCO, the Council of Europe and the European Union. Among the norms suggested by these organizations, UNESCO's declaration on the human genome and the European Council's convention on human rights and biomedicine, colloquially referred to as the Oviedo convention, are those most often mentioned. And between the two, the European Council's convention is, legally speaking, the most constraining for national governments: an international declaration creates a moral obligation, whereas a convention creates binding legal obligations, which however are rarely enforced by an international authority. In addition, the European convention, unlike UNESCO's more ambiguous declaration, contains some specific measures, one being rather controversial in several countries. In fact, the Oviedo convention prohibits the creation of embryos for research (Charles and Claeys 2001: 39–40). Although France has yet to ratify the Oviedo convention, the country's bioethics laws conform to all of its measures. It should moreover be underlined that the 1994 bioethics laws precede the adoption of both UNESCO's declaration and the Oviedo convention, both issued in 1997. In fact, the emphasis placed on the respect of human dignity in the French bioethics laws has constituted a source of inspiration for both of these transnational normative documents.

It should be said, however, that the UK's Human Fertility and Embryology Authority (HFEA) was a key model for the Socialist government when it began revising the 1994 bioethics laws in view of encouraging embryo research. Although the Agence de Biomédecine shares some similarities with the HFEA, Chirac's government was not as enthusiastic as Jospin's towards the British model.

Internationalization has had more time to influence French policies in the GMO area. As we saw above, the OECD has been developing norms in this area since the 1980s, and the European Union's first directives on GMOs date back to the early 1990s. Despite the intensive efforts by the OECD and the European Union, they offer little in terms of explanation of the French GMO policy design. It is true that in the 1980s France implemented a step-by-step approach, which had been conceived by the OECD, to authorize releases of GMOs into the environment. It should also be underlined that France made a significant contribution to the work of the OECD on this matter at the time. But more significantly, the OECD in 1989 clearly stated that the process-based approach, notably employed in France, was unnecessary, pressing countries to adopt the product-based approach (Chevassus-au-Louis 2001: 50–54). Ever since, the United States has referred frequently to this OECD norm in disputes against France over the regulation of GMOs. Despite this pressure and a reform of the country's regulatory arrangement for GMOs, French policy-makers never considered moving in the direction of the product-based approach.

Naturally, the persistence of the process-based approach might be partly explained by the European directives, which are themselves consistent with this approach. Furthermore, European law has precedence over domestic law in Europe. Therefore, one might argue that French policy-makers have no choice but to maintain this process-based approach. It should be noted, however, that the GMO directives adopted during the 1990s only served to institutionalize policies that were already prevailing in France. As argued above, Directive 90/220/CEE on the environmental release of GMOs only forced France to recognize into law the CGB and its practices. The directive did not alter in any way the work of the regulatory organization. As explained above, the European Court of Justice ruled against France with regards to the application of 2001/18/CE. Although French policy-makers are working towards a policy response to comply with the directive, decisions still have to be made.

Directive 90/220/CEE specifies a process whereby biotechnology firms might avoid having to go through the approval process of all fifteen member states in order to commercialize a GMO across Europe. Following this directive, a GMO is deemed ready for commercialization across Europe once a member state considers it safe and no other member state objects within sixty days. This procedure turned out to be instrumental for French policy-makers who, in an effort to regain a public confidence shaken by GMO opponents, sought tougher European labeling and traceability rules. Thanks to the comitology process that Directive 90/220/CEE launches when member states object to the approval of a GMO by one country, France was able to lead a coalition capable of maintaining a de facto moratorium on the commercialization of GMOs. In the case of France, internationalization did not translate into transnational pressure on domestic biotechnology policy designs. Rather it has thus far constituted an opportunity for technology-concerned actors to block the commercialization of GMOs as well as create pressure on other member states. The labeling and traceability rules pressed by France through the European Union have been, in fact, unsuccessfully resisted by the United Kingdom.


## CONCLUSION

The French ART and GMO policies correspond to intermediate designs, but for different reasons. In the ART sector, fertility treatments (i.e., the application of biotechnology) are subjected to relatively light restrictions. In contrast, embryo research is subjected to heavy restriction. In the GMO sector, the commercial use or the application of biotechnology is subjected to heavy restrictions, but policy measures related to research are more permissive. We therefore argued that the two intermediate designs require two distinct explanations.

Both sectors are characterized by distinctive network situations. In the ART sector, no cohesive policy community of targets was present to prevent the adoption of the strict measures regarding embryo research pressed by Conservative governments in 1994 and in 2004. Issue networks dominated the area and provided opponents to embryo research with a channel for its policy preferences. In contrast, a powerful policy community of targets was in place in the GMO sector to moderate the extent of the policy changes sought by a Socialist government who had to accommodate its Green coalition partner. In addition, starting in 1997, actors concerned with GMOs could count on an increasingly effective issue network to exert some influence.

If in the ART sector, clinical applications are not restricted as much as embryo research, it is largely because the Conservative governments of 1994 and 2004 did not only seek to protect embryos, but also encouraged family values. Fertility treatments for heterosexual couples were believed to serve as an encouragement to families. In the GMO sector, if the policy community of target groups could not prevent the adoption of commercial restrictions, it is largely because of the European Union. Not that the European Union itself imposed those restrictions; in fact, France never appears to be in a hurry to apply the Union's directives. Rather, GMO opponents, with the support of the coalition government, effectively used a directive dating back to 1990 to prevent commercialization across the European Union.

## NOTES

1. For a detailed analysis see Schiffino and Varone (2005).
2. Confidential interview conducted in Paris in June 2002. Translation by the authors.

## REFERENCES

Charles, B., and A. Claeys. 2001. Rapport d'information déposé en application de l'article 145 du règlement par la mission d'information commune préparatoire au projet de loi de révision des «lois bioéthiques» de juillet 1994, Tome I: rapport, *Les documents d'information de l'Assemblée nationale*, no. 3208 (tome I: rapport et tome II: auditions).

Chevassus-au-Louis, B. 2001. *OGM et agriculture: options pour l'action publique*, Commissariat Général du Plan, September 2001.

Coleman, W. D. 1998. From Protected Development to Market Liberalism: Paradigm Change in Agriculture. *Journal of European Public Policy* 5 (4): 632–51.

Engeli, I. 2004. France: Protecting Human Dignity while Encouraging Scientific Progress. In *Comparative Biomedical Policy: Governing Assisted Reproductive Technology*. Eds. I. Bleiklie, M. L. Goggin and C. Rothmayr. Pp. 138–54. London: Routledge.

Hall, P. 1986. *Governing the Economy: The Politics of State Intervention in Britain and France*. Oxford: Oxford University Press.

Héritier, A. 1999. *Policy-Making and Diversity in Europe: Escape from Deadlock*. Cambridge: Cambridge University Press.

Joly, P.-B. et al. 2000. *L'innovation controversée: le débat public sur les OGM en France*. Grenoble: INRA.

Kohler-Koch, B. 1996. Catching Up with Change: The Transformation of Governance in the European Union. *Journal of European Public Policy* 3 (3): 359–80.

Montpetit, É., C. Rothmayr, and F. Varone. 2005. Institutional Vulnerability to Social Construction: Federalism, Target Populations, and Policy Designs for Assisted Reproductive Technology. *Comparative Political Studies* 38 (2): 119–42.

Radaelli, C. 2000. Policy Transfer in the European Union: Institutional Isomorphism as a Source of Legitimacy. *Governance* 13 (1): 25–43.

Risse-Kappen, T. 1996. Exploring the Nature of the Beast: International Relations Theory and Comparative Policy Analysis Meet the European Union. *Journal of Common Market Studies* 34 (1): 53–80.

Roy, A. 2001. *Les experts face au risqué: le cas des plantes transgéniques*. Paris: Presses universitaires de France.

Schiffino, N., and F. Varone. 2005. *Régulation publique des biotechnologies: Biomédecine et OGM agroalimentaires en Belgique et en France*. Gent: Academia Press.

# 7

# ART AND GMO POLICIES IN GERMANY: EFFECTS OF MOBILIZATION, ISSUE-COUPLING, AND EUROPEANIZATION

*Gabriele Abels (Bielefeld University) and*
*Christine Rothmayr (Université de Montréal)*

According to a popular (self-)image, Germans are the people most skeptical about biotechnology and its various applications. This skepticism is reflected in the overall restrictive legal situation. Germany was among the first European countries to introduce laws regulating genetically modified organisms (GMOs) in the agro-food sector (green biotechnology)—having begun before the first EU directives were adopted. Germany is also among the early and strict regulators of red biotechnologies, in particular, assisted reproductive technology (ART). Eurobarometer data, however, reveals that support for biotechnology applications is actually not weaker than in other countries; in the case of red applications Germans are mostly supportive (see Hampel et al. 1998; Gaskell et al. 2003; Eurobarometer 58.0). How can we thus explain the comparatively very restrictive regulation of ART and GMOs in Germany?

The social and political debates on ART and GMOs were closely linked in Germany. Since the early 1980s, both topics were widely and critically discussed in a "programmatic debate" on the impact of technological progress on society. In this highly politicized and ideological debate, we observed a clash of opposing views of the world and nature ("Welt- und Naturbilder'"; Behrens et al. 1997: 43f.; Gill 2003). The Left's strong political mobilization against biotechnology—supported by religious interest groups and the churches in the case of ART—demanded active intervention from the state in order to protect human dignity and the right to live; in the case of GMOs, the main focus was on environmental protection and public health. The multiple access and veto points of the German political system allowed the proponents of strong state intervention to successfully lobby for restrictive policies.

While the linkage of the two sectors might help to explain parallels in actor mobilization, differences in the degree of Europeanization pose an interesting puzzle: in the ART field national legislation is dominant, whereas EU legislation dominates GMO regulation. Hence we need to explain the similarly restrictive outcomes despite the different impact of European integration on national policy-making in the two fields. To do so, we present the two cases and compare them with regard to the explanatory variables outlined in chapter 1.

## RESTRICTIVE REGULATIONS FOR ART TREATMENT AND EMBRYO RESEARCH

The designing process can be divided roughly into two sequences resulting in two major acts:[1] the *Gesetz zum Schutz von Embryonen* (*Embryonenschutzgesetz–EschG*, Embryo Protection Act, 1990) and the *Gesetz zur Sicherstellung des Embryonenschutzes im Zusammenhang mit Einfuhr und Verwendung menschlicher embryonaler Stammzellen* (*Stammzellgesetz– StZG*, Stem Cell Act, 2002). Biotechnology in the human and nonhuman field has been on the political agenda since the early 1980s. In 1985, the *Bundesärztekammer*, the German Medical Association, issued the first binding decision on ART ("Guidelines on Assisted Reproduction") in 1985.[2] Simultaneously, the first concrete recommendations for legal measures were elaborated by the Benda Commission, a joint expert group of the Federal Ministry of Justice and the Federal Ministry of Research and Technology. The Benda Report (1985) led to an embryo protection bill proposed by the Ministry of Justice in 1986. As the *Länder* are important players in health care provision and the supervision of the medical profession, they also began concurrently to engage in the ART debate. The *Bundesrat* (upper chamber composed of representatives of the *Länder* governments) decided in 1986 that the *Länder* together with the Bund should elaborate a comprehensive legislative concept (BR-Drs. 210/86) resulting in the establishment of a common working group of the federal government and the *Länder*. However, the Ministry of Justice's second Embryo Protection Bill (1988), introduced into Parliament in 1989, did not follow the report of the common working group (Abschlussbericht Bund-Länder-Arbeitsgruppe of

1988) that had recommended comprehensive ART legislation through a combination of federal criminal law and *Länder* laws. The Ministry of Justice's bill passed the *Bundestag* (lower chamber of elected members), with the clear majority of the governmental coalition of Christian Democrats (CDU/CSU) and Liberals (FDP) despite propositions for more comprehensive federal legislation and even more restrictive regulations proposed by the opposition Social Democrats (SPD) and Greens.

ART became once again a contested issue at the end of the 1990s. The renewed debate now focused on the question of the protection of human life against the background of new developments in the field of biomedicine. It was influenced by the first cloning experiments and the speech of the German philosopher Peter Sloterdijk in 1999 on *Regeln für den Menschenpark*, or "rules for the human park" (see Geyer 2001; Graumann 2001). Embryonic stem cell research was at the core of the debate (see Gottweis 2002), leading to the 2002 StZG. In fact, the Bundestag itself took the crucial decision to put the issue on the political agenda again by establishing the parliamentary *Enquetekommission Recht und Ethik der modernen Medizin* in March 2000 (Parliamentary Inquiry Commission on Law and Ethics in Modern Medicine, BT-Plenarprotokoll 14/96). Based on the commission's report, three different motions on stem cell research were introduced to Parliament (BT-Drs. 14/8101, 14/8102, 14/8103): a complete ban; a change of the EschG in order to allow research *and* derivation of stem cells in Germany; and a third motion allowing only research on imported embryonic stem cells under strict conditions. The last motion prevailed (BT-Plenarprotokoll 14/214: 21239-61) leading to the StZG. The debate, however, continued after the adoption of the StZG. In mid-May 2005, Chancellor Schröder declared a "change in biopolitics" towards a more research-friendly solution in the long run, which would even permit therapeutic cloning. The SPD lost the election in North Rhine-Westphalia shortly thereafter, resulting in an additional conservative *Länder* government and, thus, further strengthening the conservative majority in the *Bundesrat*. As a result, Schröder unexpectedly announced early general elections in September 2005. Researchers feared that a new conservative government—which seemed very likely at the time—would favor a restrictive interpretation of the StZG instead of adapting a more researcher-friendly solution (Bahnsen and Willmann, 2005: 29f.). The election, however, resulted in a "grand coalition" government of CDU/CSU and SPD. The coalition agreement states that biotechnology in the pharmaceutical sectors will be supported. Regarding stem cell research, adult stem cell research is considered to be very important. A number of underregulated and controversial issues, however, are not mentioned (e.g., embryo research, preimplantation diagnosis or genetic testing).

The German ART policy design corresponds to the restrictive ideal type. In order to protect the embryo, human dignity and the right to life, current regulations strongly restrict ART treatment and research. The EschG, which declares certain techniques and practices to be criminal offences, almost exclusively governs ART. It prohibits egg donation, embryo donation, preimplantation diagnostics, and the transfer of more than three embryos to a woman within one cycle. Furthermore, the creation of an embryo for any purpose other than inducing a pregnancy is not allowed, thereby outlawing any research on embryos and totipotent cells. In line with this restrictive framework, the StZG stipulates exceptions under which a research project might use imported stem cells, while leaving some gaps (e.g., within research on stem cell products).

Because of their legally binding character as part of the medical professional code, the "Guidelines on Assisted Reproduction" of the *Bundesärztekammer* (German Medical Association), last updated in 1998, are also part of the policy design. Through a combination of quality standards, information and counseling requirements, reporting and documentation duties, as well as various regulations on how and when to practice and what type of technique, the guidelines further regulate ART treatment (see table 7.1). Additionally, the eligibility for ART treatment is limited to married couples. Stable unmarried couples are admitted on a case-by-case basis. Coverage by the public health care insurance is granted for a range of techniques. Yet, since 2004 public health insurance providers pay only for 50 percent of the costs for up to three cycles; in addition, coverage depends on marital status and age. This leads to the development of a small IVF market and presumably also to a drop in the number of IVF cycles per year in Germany.

The main implementers of these policies are the *Länder Aerztekammern* (Länder Medical Associations) and the *Bundesministerium für Gesundheit* (Ministry of Public Health). The StZG is implemented by the Ministry of Public Health and a newly instituted *Zentrale Ethik-Kommission für Stammzellenforschung* (Central Ethics Commission for Stem Cell Research).

## RESTRICTIVE REGULATION OF GM CROPS AND FOOD

The policy design process for agro-food biotechnology developed in four stages: Between 1975 and 1984, self-regulation of researchers and produc-

**Table 7.1.    Current German ART Policy Design**

| | Research | Practice of fertility treatments | Measures aimed at patients |
|---|---|---|---|
| Policy-Design | Creation of embryos, embryo research, cloning and derivation of stem cell lines prohibited<br><br>Research on imported stem cell lines permitted under specific conditions, authorization required | PID, embryo, and egg donation prohibited; transfer of embryos limited to three | Access limited to married couples, on a case-by-case basis to stable couples; costs only partly covered; no PGD allowed |
| Type | Restrictive | Restrictive | Restrictive |

ers was the dominant design. A first initiative for a special law on genetic engineering failed in the 1970s (in detail Waldkirch 2004). During the period from 1984 to 1990, national legislation came to the fore. A first public debate led to the institution of the parliamentary *Enquetekommission Chancen und Risiken der Gentechnologie* (Inquiry Commission "Chances and Risks of Genetic Engineering") in 1984. The commission and its 1987 report stimulated further public debate. The commission's majority recommended the development of a legal framework based on existing guidelines. Spurred by national debates and new legislation, a discussion began about European rules, which, it was becoming quickly quite obvious, would take a fairly restrictive process-based approach (cf. chapter 2 in this book). The need for a reliable legal basis increased when the pharmaceutical company Hoechst filed legal action against the Hessian government, because its application for licensing of a production site for GM human insulin had been rejected. The Hessian administrative court (Hessischer Verwaltungsgerichthof) sustained the rejection arguing that the existing legal basis (Federal Emission Protection Act) for licensing was insufficient. These developments created additional pressure—and electrified the German industry that was starting to develop a strong interest in biotechnology—for what then became the 1990 *Gesetz zur Regelung der Gentechnik* (GenTG, Act on Genetic Engineering), which aimed at promoting biotechnology while protecting people and the environment from risks. This law had already adopted a process-based approach (i.e., genetic engineering was distinct and, therefore, required unique legal regulation). The *Robert Koch*

*Institut* (RKI) became the competent authority for authorization, assisted by the newly established *Zentrale Kommission für biologische Sicherheit* (ZKBS, Central Commission for Biological Safety).

The third stage, from 1991 to 1997, was characterized by continued national debate and, above all, the Europeanization of GMO regulation. The EU adopted the so-called Contained Use Directive (90/219/EEC) and the Deliberate Release Directive (90/220/EEC) in 1990. Since EU law supersedes national law, revision of the GenTG became necessary in order to transpose these EU directives, a process deeply involving the *Länder* as the main implementers (Bandelow 1997). The 1993 revision of the GenTG responded to the criticisms of researchers and industry. Supported by the government coalition as well as the SPD, the GenTG now emphasized the need to promote the economic potential of biotechnology (see Waldkirch 2004: 169f, 191). For example, the safety standards for certain laboratories working with GMOs were deregulated and application procedures were simplified. The RKI was nominated as the national authority responsible for dealing with applications for field trials of GM crops, based on Directive 90/220/EC. An intended further deregulation of the German GenTG was not possible, because of the stricter EU law (Waldkirch 2004: 172).

A debate about a special EU law covering GM food had already begun by the early 1990s. Galvanized by the spread of the BSE crisis and the failure of regulators to manage it effectively, food safety became a highly political issue in the mid-1990s. In this distrustful climate, US biotech companies like Monsanto forcefully tried to import GM soybeans and maize to the European internal market in 1996–1997. From that point on, GM crops and especially GM foods became the focus of public debate (see Lassen et al. 2002) and sectoral interest groups engaged in "campaign politics" (see Behrens et al. 1997). The conservative-liberal German government now focused on supranational regulation, as there was a political impetus to fill in the existing regulatory gaps (e.g., GM food, labeling). This phase lasted until the EU adopted "Regulation No. 258/97 on Novel Foods and Novel Food Ingredients" in 1997 adding restrictive product-specific requirements to the existing process-based approach. GMOs intended for use in food now required case-by-case approval and labeling requirements to allow for consumer choice. On the national level, the Robert Koch Institute (RKI) was placed in charge of approving GM food, yet depending on the type of product, the *Bundesinstitut für Gesundheitlichen Verbraucherschutz und Veterinärmedizin* (BgVV, Federal Institute for Health Protection of Consumers and Veterinary Medicine) were also to be involved.

A fourth phase (1998 to 2004) was characterized by substantial and institutional changes at the EU and national level. Triggered by the general food safety crisis, public outrage, and due to massive conflicts in the implementation committee responsible for authorizing GMOs between EU member states, the environmental council adopted in 1998 a de facto moratorium on market releases of GMOs. This moratorium created enormous pressure to revise the legislative framework and the Deliberate Release Directive was revised in 2001. The new Directive 2001/18/EC tightened the rules for authorization. It introduced a ten-year time limit for first time authorization, abolished the simplified notification procedure, and established a comprehensive monitoring system. As a consequence, the German GenTG had to be revised again. In 2002, the transposition of 98/81/EC (i.e., the revised Contained Use Directive) into the German GenTG introduced safety standards more restrictive than those required by EU minimum standards. The transposition of Directive 2001/18/EC led to a similar outcome, by reviving the fundamental conflict over the "coexistence" of GMO and GM-free agriculture. Given that there is no EU law on this critical issue, member states have room for maneuver. A central issue of the debate was the notion of strict liability that held farmers responsible irrespective of individual wrongdoing (*verschuldungsunabhängige Haftung*). Fearing high economic costs, producers and farmers demanded a public fund for compensation, yet, the minister and environmental groups opposed it. When the conservative majority in the *Bundesrat* threatened to veto the bill, the federal government split it into a general part, which did not require *Bundesrat* approval, and a second part on technical and administrative implementation (e.g., minimum distances between fields) needing approval. The former, highly stringent bill then passed the Bundestag in November 2004 and has been in effect since February 2005. In response, the *Deutsche Bauernverband* (German Farmers Association) advised its members not to grow GM crops in order to avoid possible litigation claims. Even before notification, in July 2004, the EU Commission released a position paper announcing that the government bill would introduce "through the back door" national norms that were far too stringent and irreconcilable with European law.

The conservative majority in the *Bundesrat*, together with some SPD *Länder* governments, accused the new GenTG to be a *Gentechnikverhinderungsgesetz* (law to prevent genetic engineering). In April 2005, they voted against the federal government's bill for implementation provisions and transferred the bill to the mediation committee between the two chambers; the committee was unable to reach a compromise. Using their veto power, this coalition tried to put pressure on the government to introduce changes to the GenTG. The coalition opposed in particular the liability rules and public access to registers on field trials; they preferred a more farmer- and innovation-friendly regulation of liability. In the case of public access to registers on field trials, they feared that opponents of biotechnology would use them to locate individual fields. Furthermore, the conservative government of Saxony-Anhalt decided to challenge in April 2005 the constitutionality (*Normenkontrollklage*) of the new GenTG at the *Bundesverfassungsgericht* (German Supreme Court) since it violated the right to freely choose one's occupation or profession, the right to property, the freedom of research as well as the principle of equality before the law, all of which are guaranteed by the *Grundgesetz* (GG, German Basic Law).

Agro-food biotechnology was politicized as a key field for economic innovation in the 2005 election campaign. Conservatives and Liberals promised to revise the GenTG to make it more "innovation-friendly." The grand coalition that is currently forming the new government is more divided on green biotechnology. The compromise reached in the agreement on the coalition states that the GenTG will be revised again, although regulatory goals remain the same (i.e., insure choice for farmers and consumers and render vague concepts more precise). With regard to the controversial liability rule, the government wants a compensation fund to be set up by the relevant branches of industry. In the long run, an insurance model would be preferred to the compensation fund. In addition, the law will be altered to recognize both the principle of liability for fault only, in case a farmer violates professional rules (*Regeln der guten fachlichen Praxis*) and the precautionary principle. However, these rules must still be established, thus providing a new potential arena for conflict.

With respect to GM food, directly effective EU legislation was introduced in 2003. Regulation (EC) No. 1829/2003 on GM food and feed as well as Regulation No. 1830/2003 on traceability and labeling installed a common framework for GM food and feed as part of the new food safety regime. Following these regulations, the 1998 moratorium was finally lifted and yearlong pending authorizations were finally approved. The regulations introduced stricter standards (e.g., for labeling or monitoring). Furthermore, a new centralized authorization procedure created links between the European Food Safety Authority and national competent authorities such as the *Bundesinstitut für Risikobewertung und –kommunikation* (Federal Institute for Risk Assessment and Risk Communication), which was responsible for scientific aspects, and the *Bundesamt für Verbraucherschutz und Lebensmittelsicherheit* (BVL, Federal Agency for

Consumer Protection and Food Safety), which was in charge of risk management; both were established in 2002 as part of the European Rapid Alert System.

The implementation practice in Germany is restrictive. For example, EU legislation includes a safeguard clause allowing member states to take temporary preventive measures to protect health and the environment, if there are reasonable concerns. By July 2000, Germany was one out of five EU member states (Austria, Luxembourg, France and Greece) that had invoked the safeguard clause in sixteen cases to impose bans on importing and planting GMO (e.g., Novartis Bt-corn). New crops require the authorization of the *Bundessortenamt* (Federal Agency for Seeds) in accordance with the RKI. In this case, the Federal Ministry of Health withdrew the RKI authorization and thereby made a "variety recognition" judicially impossible. Cultivation for experimental purposes, however, is permitted.[3] In addition, since 1998 GM-free foods may be labeled as "without gene technology."

In sum, GMO regulation is process-based and, in the food sector, accompanied by vertical, sectoral rules (see table 7.2). It combines strict premarket authorization and postmarket control instruments such as labeling requirements, monitoring systems (e.g., registers) and sanctions. The overall goal is to protect consumers and the environment, while simultaneously promoting biotechnology.

**Table 7.2. Current German GMO Policy Design**

|  | Field trials | Commercialization and cultivation | Consumer and environment |
|---|---|---|---|
| Policy Design | EU deliberate release directive transposed into GenTG: <br>- authorization required by competent authorities in the Länder <br>- Inspections <br>- Penalties | EU law for GM food and feed as well as deliberate release; de facto moratorium on marketing of new GMOs: <br>- Labeling requirements <br>- Authorized GMOs subject to close monitoring <br>- Liability for damages | EU law for GM food and feed as well as deliberate release: <br>- Labeling requirements defined at EU level <br>- Traceability requirements defined at EU level |
| Type | Restrictive | Restrictive | Restrictive |

## COMPARING ART AND GMO POLICIES: DIFFERENT DEGREE OF EUROPEANIZATION, SIMILAR OUTCOME

German biopolicies in both fields correspond to the restrictive type of design. National legislation strongly limits ART treatment and research. The policy design of GMOs is dominated by European legislation; implementation on the national level is, however, more restrictive than in other EU countries. We argue that the key factors explaining the restrictive design are the public mobilization of interest groups and the early coupling of the two issues, as well as the subsequent influence of these two factors on the national decision-making process in the case of ART and the implementation of EU norms at the national level for GMOs.

The German political system is characterized by "cooperative federalism" and corporatism, which both contribute to a comparably consensual decision-making style (Lijphart 1999: 247f.). This provides interest groups critical towards biotechnology with multiple access points to the policymaking process (e.g., via political parties or corporatist networks). In addition, some actors (e.g., Bundesrat) have a veto power. The need to compromise explains why neither opponents nor proponents fully realized their goals. In the case of ART, for example, the need to compromise explains why stem cell research is possible despite the strong protections on the embryo. In the GMO case, negotiations over regulation involved not only the EU and the federal level—represented from the early 1980s until 1998 by a conservative-liberal government—but also the *Länder*, which promoted a general regulatory framework, but differed along party lines with respect to the stringency of the regulations.

## PUBLIC MOBILIZATION AND ISSUE COUPLING

Policy networks in both fields are similar in terms of actor mobilization. Interest groups with concerns about biotechnologies succeeded in mobilizing the public. In the case of ART, consumer and environmental groups as well as religious and women's groups campaigned for restrictive policies. Patient groups, in contrast, played a marginal role. Green and feminist critics were generally mistrustful of science and feared ART's potential for eugenic purposes—a particularly strong concern given Germany's history and the crimes committed under the National Socialist regime. Environmentalists and feminists began organizing against ART as early as 1985 (Betta 1995: 115). ART was thus framed as part of an overall debate on biotechnology and its potentially negative impact on society. Disability rights groups and the churches also strongly advocated for restrictive policies. The former groups shared eugenic concerns, whereas the latter sought to protect unborn life. Medical and research interests were also active in the debate. The medical profession, principally the *Bundesärztekammer* (German Medical Association), did not advocate permissive policies and early on placed restrictions on ART through self-regulation. Researchers, represented by the *Deutsche Forschungsgemeinschaft* (DFG, German Research Council) and the Max Planck Society, pursued a comparably more permissive policy with respect to embryo and stem cell research, but did not promote a permissive design. Given the strong mobilization of anti-biotechnology interest groups, the network was not dominated by medical and research expert knowledge and interests.

The ART debate involved a conflict over values, whereas the GMO debate was framed in terms of economic versus public interests. Public interest groups such as environmentalists and consumer activists adopted a consistent position against biotechnology, campaigned in unity and successfully made use of public protests, especially consumer boycotts (see Behrens et al. 1997; Hampel et al. 1998). In contrast, producers (farmers and manufacturers) and retailers were divided. In the end, the alliance of economic actors fell apart and producers as well as retailers aligned with the position of the anti-biotechnology groups that supported mandatory labeling requirements (Epp 2003: 136).

The strong parallels in actor mobilization are to some extent due to the larger critical discourse on the impact of biotechnology on society, which began as early as the 1980s. The debates mobilized strong resistance from the left of the political spectrum leading to demands for strict control or often even for total bans. This successful mobilization along with direct lobbying had a strong influence on the adoption of restrictive policies. In the case of the GM agro-food sector, the lobbying did not only influence the position of the German government on the European level, but contributed to restrictive domestic implementation of EU norms, especially after a government coalition formed by the SPD and Greens came into power in 1998. Public outrage led to a de facto moratorium in the EU and resulted also in the decline of applications for experimental field release in Germany that were not directly affected by the legislation. In the case of ART, the mobilization clearly contributed to pushing the propositions for rather moderate state intervention, which had been formulated by the Benda Report, into the much more restrictive propositions in the subsequently adopted ESchG, and hence contributed to the medical and research interests not realizing their policy goals.

Public mobilization also contributed to the use of self-regulation in both fields. First of all, it promoted the adoption of relatively strict and early self-regulation by the German Medical Association. Furthermore, the DFG and the Max Planck Society, under public pressure, had to declare a moratorium on embryo research as early as 1988 (Betta 1995: 102), before the adoption of the EschG. Early self-regulation, however, neither prevented strong state intervention nor contributed to regulations in line with the interests of the medical and research community. The same is true for the GMO field. Self-regulation still prevented state intervention in the 1970s; yet from the 1980s onward, there was a broad consensus about the need for federal and, increasingly, for EU laws. Since legal regulation did not solve the political conflict, producers and retailers turned to *additional* self-regulation such as contract farming and certification that allowed for the segregation between conventional and GM products, or to positive labeling of GMO-free foods.

The nonsuccess of target groups and of self-regulation was similarly linked to the framing of the issues at stake through the mobilized public interest groups. In the GMO case, competing frames of producers/researchers versus consumers/environmentalists clashed. In the related EU regulation, GM food was defined as "novel food" meaning food not widely used at that point in the EU. Yet this framing emphasizes the cognitive limits since "novel" implies a lack of experience based on (scientific) evidence; there is no final scientific proof of safety and new scientific research may modify prior risk assessments (Epp 2003). As a result, consumers did not consider scientific risk assessment and associated regulatory agencies as trustworthy—especially given the instrumental use of scientific knowledge in the handling of the BSE crisis. Neither did they trust industrial actors. GM food was normatively perceived as insufficiently safe. Producers then became afraid of a negative image and of consumer boycotts. With the policy changes introduced at the EU level, GM food and feed then became a separate category emphasizing the process-based regulatory approach. In so doing, GM food policy followed the route of process-based GMO regulation.

In the ART field, two types of framings promoted restrictive solutions. ART was framed in terms of a potentially risky new technology and as a constitutional question of embryo protection, and not as a medical treatment for infertile couples or a health care issue. The German Supreme Court's decisions on abortion law in 1975 declared that the constitutional protection of human dignity (Art. 1 of the German Basic Law, GG) and the right to life (Art. 2 GG) applies to the embryo after implantation (Wilde 2001: 183–85). The state accordingly recognized the obligation to protect the embryo also with regard to research. This jurisprudence strongly contributed to the framing of ART and worked in favor of the proponents of restrictive policies with regard to the EschG and the StZG.

The media played an interesting role in the biotechnology debate in Germany. Mediatization theorists claim that the media have an almost determinative role on the thematization of public issues and public opinion. Politicians and industry often argued that the media's negative representation of biotechnology was responsible for negative public opinion. However, in-depth analysis of print media clearly illustrates that there has been a *balanced* and overall even *positive* representation of biotechnology (Hampel et al. 1998; see also Hampel and Renn 2001; Kepplinger 1995; Graumann 2002). Therefore, the media does neither reflect nor determine public opinion, but it does influence the public agenda.[4] It is, however, clear that when events became mediatized, public opinion became more critical, for example, in the field of biomedicine, this occurred following the reports on attempts of human cloning and the Sloterdijk debate, and in the agro-food sector especially after the first importation of GM soybeans from the US to Europe. In summary, we do see a mediatization of the debate in both fields, but the linkage between media representation (at least with regard to the print media), public opinion and public policy is not straightforward. Thus, we cannot make any claims about the effects of the media debate on the resulting policy design.

## PARTY POLITICS, FEDERALISM, AND THE ROLE OF ADMINISTRATION: RESTRICTIVE REGULATION AS COMPROMISE

Regulations in the two fields differ in terms of the influence of political parties and the Bundestag. In the ART field, domestic party politics and the Bundestag played an important role because of the lack of supranational norms. This was less so in the GMO field due to Europeanization, which obstructed the traditional arena for party politics. Transposition of EU regulations into national laws revitalized, on occasion, the conflict along party lines (e.g., the 2004 revision of the GenTG transposing EU Directive 2001/18/EC). The EU is increasingly using regulations (i.e., directly effective legislation) and thereby further marginalizes the *Bundestag* and, thus, a route for party politics. This said, we are still able to compare how government and opposition influenced the final design for ART and GMO in the agro-food sector.

The Green Party played an important role in both fields, as an opposition party, and from 1998 to 2005 as a government party, contributing to the restrictive design of the ESchG, and the stringent transposition and implementation of EU norms regulating GMOs. The party manifesto is extremely critical of the social effects of agro-food biotechnology, referring to the ecological and health risks, as well as the increasing dependence of farmers on agro-business. In 2001, all responsibility for GMOs was turned over from the Federal Ministry of Health to the newly founded *Bundesministerium für Verbraucherschutz, Ernährung und Landwirtschaft* (Federal Ministry of Consumer Protection) headed by a Green Minister, Renate Künast, thus increasing the influence of the Greens. For example, Künast stopped the authorization procedure for the GMO crop "Artuis" in June 2001 and was skeptical about biosafety research on GMOs promoted by the *Bundesministerium für Bildung und Forschung* in 2004 (Federal Ministry for Research); in 2005 the Ministry for Consumer Protection started its own research on GM maize and technical options for securing coexistence. Regarding ART, the Greens, when adopting the EschG in opposition and the StZG in government, saw itself as a "critical companion." The three main criteria for evaluating reproductive medicine and stem cell research were human dignity, which supported strong protection of the human embryo and of women, civil rights, and the diversity of human life. The second major party in the ART field was the CDU/CSU, for whom abortion and their strong tradition of protecting the unborn life were the key issues and who was in government at the time of adopting the EschG. Given that the CDU/CSU is now part of the "grand coalition" government, changing the EschG and the StZG into a less restrictive direction (e.g., to allow for therapeutic cloning) is unlikely.

Furthermore, the comparison reveals a basic consensus building among political parties for generally strict policies. No political party advocated permissive policies. The Liberals, who were interested in promoting more permissive policies for instance in stem cell research or marketing of GMOs, are the sole exception. There were, however, differences among parties—as well as within parties—regarding the question of how restrictive policies should be. The traditional left-right divide played out more in the case of GMOs, whereas, in the case of ART, both the left and the Christian Democrats advocated restrictive to very restrictive policies. In fact, the SPD and the Greens promoted even more restrictions than the Christian Democrats. In the stem cell debate, on the contrary, opinions within parties were divided, especially in the SPD. In the GMO field, the left has traditionally been more skeptical and in favor of more restrictive policy designs, whereas the CDU/CSU and the FDP strongly favored agro-food biotechnology as a key technology for future economic development. However, there is also strong support for an economic rationale in parts of the current SPD—a fraction led by the chancellor himself, which has led to some conflicts with the Greens, the junior partner in government and the one in charge of biotechnology policy (see above). Yet, not all party members support the "modernization route" of the party leaders. The current constellation does not, therefore, seem likely to produce a radical change such as introducing an intermediate policy design.

Despite an overall agreement on strict regulation in both fields, compromises needed to be forged between advocates of very restrictive policies and more moderate solutions. While in the case of the EschG, the government (CDU/CSU and FDP) and opposition (SPD and Greens) both favored restrictive policies, opinions within parties were much more divided for embryonic stem cell research. In fact, the parties decided to let their representatives in the Bundestag vote freely on the issue; this opened the door for the compromise between proponents of total prohibitions and advocates of research, which allowed for research on imported stem cell lines under certain conditions. An additional institutional characteristic relevant to the field of ART is the German "Chancellor Democracy" (Niclauss 1999: 37). Chancellor Schröder took the initiative in the stem cell debate by publicly taking a liberal position on embryonic stem cell research and establishing the National Ethics Council in order to create a counterweight to the initiative of the Bundestag in establishing an *Enquetekommission*. However, he did not succeed in realizing his policy goal of even allowing the derivation of embryonic stem cells for research in Germany. This "power play" illustrates well the German system of separation of powers as well as the multiple points of veto or access, enhancing, in the case of ART, the search for a viable consensus.

Another element of "separation of powers" is the bicameral legislature and its different modes of composition. Different parties regularly dominate the *Bundestag* and the *Bundesrat*. This German form of "cohabitation," institutionalized via the mediation committee, is one factor that enhances a consensual style of policy-making (e.g., Lijphart 1999: 248). In 1990, the *Bundestag* was dominated by the conservative-liberal majority while the *Bundesrat* was dominated by the SPD. When the most comprehensive revision of the GenTG took place in 2004, SPD and Greens were in government holding a majority in the *Bundestag* and facing a strong conservative majority in the *Bundesrat*. "Divided" government demands compromises. In the case of the 2004 GenTG revision, the bill was split in order to circumvent the *Bundesrat's* opposition to the governmental proposition. This bill then became law, but is now facing additional critiques from the European Commission, as well as from several *Länder* governments. Whether or not policies for green and red biotechnology will diverge in the future is currently unclear. A CDU/CSU-led government would likely revise the GenTG within the boundaries set by EU legislation in order to make it more innovation-friendly (e.g., abolishing the strict liability rules that are claimed to be a de facto ban of GMOs). Yet, the junior partner SPD would have to agree to change the law, which had previously been adopted by the former SPD/Green government. A revision of ART regulation, especially the StZG, in order to foster researchers' interests is highly unlikely.

Furthermore, the German style of cooperative federalism influenced the decision-making processes in both fields. There was no competition between the federal level and the *Länder* or between the *Länder*; in fact, there was cooperation in order to adopt national norms or to influence European policies. The difference between the two fields lies in the importance of the *Länder* and the *Bundesrat*, as well as the policies advocated by the majority of the *Länder*. We have to take into account that differences between federal and *Länder* level often coincide with differences between parties. Because the *Bundesrat* consists of *Länder* governments, its composition changes frequently due to election dates being different on the *Länder* and federal levels. As a result, the majority in the *Bundestag* constituting the federal government often faces a situation where the minority party in the *Bundestag* is the majority party in the *Bundesrat*. Thus, party politics is interlinked with multilevel governance.

Although the *Länder* had considerable responsibilities in both fields, they were more important in the ART field, because of the lack of Europeanization in ART policy-making. In principle, it would have been possible to adopt *Länder* laws, and some *Länder* elaborated drafts. German federalism, however, corresponds to the model of a "unitary federal state": cooperation between *Länder* and *Bund* as well as cooperation between *Länder* aims at establishing nationwide equal legal, economic and general living conditions (Benz 1999: 136). The *Länder*, in fact, exerted pressure to adopt *national* ART regulation and overwhelmingly supported restrictive policies. In the case of GMO, opinions between the *Länder* were more divided along party lines. Some *Länder* favored restrictive policies (e.g. the government of Schleswig-Holstein, a coalition of the SPD and Greens, who introduced a *Bundesrat* bill for an amended GenTG generally supportive of the restrictive Künast proposal). Many *Länder*, especially but not exclusively conservative ones, were in favor of a less restrictive transposition of the EU legislation in 2004 into the GenTG and wanted to promote GMOs for economic reasons. While the *Länder* influence differed between the two fields, it was more important and pushed in a more restrictive direction in the case of ART than GMO. We observed in both fields a shift of competences to the federal level, but for different political reasons; in the case of ART, the political reasons played out on the *Länder* and federal level, whereas in the case of GMOs European integration played a more important role. In both fields, the *Länder* remain key players in the implementation, yet more so in the field of ART where authorization and control of the practice are left to the *Länder* and the *Länder* Medical Associations (with the exception of stem cell research). Regarding GMOs, the federal level is in charge of authorization but the introduction of a centralized European procedure involving national competent authorities reduced the importance of the *Länder* except regarding the authorization for contained use and experimental field release.

A further variable to consider is the role of ministries. The findings support the hypothesis that the type of ministry responsible and the political party in charge of the ministry contribute to explaining the restrictive policies. In the ART field, the ministries involved, such as the Federal Ministry of Justice, but also ministries for research and of public health, coordinated the elaboration of the EschG and created deliberative arenas by establishing expert commissions and working groups. The dominant position of the *Bundesministerium der Justiz* (Federal Ministry of Justice) certainly had an impact on how embryo research was approached; the use of decisions of the *Bundesverfassungsgericht* on abortion reinforced the constitutionalization of the debate. Hence, the strong focus on embryo protection contributed to restrictive policies.

However, we need to consider that the *Bundestag* gained influence by establishing the *Enquetekommission*. Reinforced by a lack of cohesion in the governmental and opposition parties, the parliamentary arena was the dominant one in elaborating the StZG. In fact, it was elaborated on the initiative and under the lead of the *Bundestag* and the *Enquetekomission* and not a specific ministry. The chancellor did not succeed in creating a counterweight to this initiative by publicly taking a liberal position and establishing the *Nationaler Ethikrat* attached to the *Kanzleramt* (National Ethic Commission attached to the Chancellor's Office).

In the GMO field, power plays took place between different federal ministries rather than between the Chancellor and the *Bundestag*. In the 1990s, the Federal Ministry of Public Health was responsible for GMO policy; in 2001 the Ministry for Consumer Protection took over. The Green Minister then tried to induce changes in the policy network when she initiated a deliberative exercise, *Diskurs Grüne Gentechnik*, involving all stakeholders and thereby creating a "window of opportunity" (see Kingdon 2003) for opponents of GMOs. This was part of the ministry's work on a proposal for transposition of the EU deliberate release directive into the GenTG. Four SPD-led federal ministries (justice, economy, research and public health) were involved in the policy process—all supporting the chancellor's position that biotechnology is essential for economic competitiveness and industrial modernization. The fact that the Green proposal prevailed and that the government even split the bill to avoid the potential strong influence of the CDU/CSU and FDP via their *Bundesrat* majority reveals the importance of the issue at hand. A plausible interpretation is that the SPD wanted to avoid a fight within their own party as well as with the Greens and instead rely on the effects of multilevel governance (i.e., intervention from the EU level as well as from the *Länder* via the *Bundesrat*).

## EUROPEANIZATION OF NATIONAL LAW AND POLICY TRANSFER

The two fields differ considerably regarding globalization and Europeanization. While there is no such influence in the ART field, Europeanization (and globalization) is the central factor in the GMO case. In the course of the 1990s, GMO regulation became Europeanized, thus superseding national law and resulting in institutional change at the national level. EU law, in the form of regulations, is increasingly directly effective. When issued as a directive, European law has to be transposed into national laws, but this leaves member states some discretionary freedom over implementation. Since the mid-1990s, EU legislation on GMOs has become increasingly restrictive (see Abels in this book) and has thus led to a restrictive implementation and transposition of EU norms in Germany. Yet, the latest GenTG revision is even more restrictive than required. In order to explain national transposition and implementation practice, we referred to actor networks and the characteristics of the German political system. One of the problems anti-biotechnology groups encountered was that the Europeanization of GMO regulation made public mobilization harder and more resource-intensive. In general, the EU is a responsive institutional environment for public interest groups, because the balancing of private economic interests by including "diffuse interests" helps to increase the overall weak legitimacy of the EU (see Pollack 1997). Skogstad (2003) argues that in the field of GMOs "network governance" has been most important to enhance the legitimacy of EU policies. However, Europeanization of rule-making makes national mobilization and lobbying anything but obsolete. There is rather a division of labor regarding lobbying efforts at the national (e.g., the German BUND) and EU levels (e.g., Greenpeace, Friends of the Earth Europe). On the one hand, mobilization at the national level has clearly contributed to an increasingly restrictive policy design at the European level; on the other hand, the transposition of European policies into national laws allows for the reopening of the national arena (see the 2004 revision of the German GenTG).

Market factors are, in general, highly influential in regulation. Within Germany, however, market competition was not a predominant factor for explaining the policy outcome. Producer interests such as seed producers (e.g., *Kleinwanzlebener Saatzucht*) or chemical companies (e.g., Bayer-CropSciences or BASF) were involved in the policy-making process, but on a global scale they were not as strong players as companies such as Monsanto. German companies have been generally latecomers to this industry. Today, GM crops are grown on less than 0.05 million hectares in Germany (to 47.6 million hectares in the United States; see James 2004). Even the number of field experiments has decreased over the years: between 1991 and 2002 only 120 field trials were signaled to the EU Commission (compared to an average of fifteen hundred field trials carried out annually in the US; Rapid Press Release April 17, 2002). At the same time, there were 228 applications for deliberate releases at 643 geographical locations in Germany (12 percent of the EU total). It is above all the strong public resistance to deliberate releases of GMOs in general and GM food in particular that explains the policy outcome along with the lack of strong economic interest groups. German biotech companies in the agro-food sector invest now into second generations of GM crops, which promise advantages for consumers, and, thereby, hope to decrease consumer resistance. The Swiss referendum of November 2005, however, accepting a five-year moratorium for GM plants and animals in agriculture, contributed to once again stirring up the public debate and creating resistance towards GM crops and food in Germany and other European countries.

With respect to ART, international research competition strengthened the position of research interests and indirectly contributed to leaving the door slightly open for embryonic stem cell research in Germany. There is also some bottom-up influence from the national to the international level: the EU guidelines for embryonic stem cell research (see Abels in this book) were modeled after the German policy introducing a deadline for the use of donated embryos. Yet, for other aspects of ART globalization has not been an important issue due to the fact that markets are national or regional, if we can speak of a market at all, since in Germany the provision of ART treatments is highly regulated via the national health system.[5] Germany has so far neither signed nor ratified the Council of Europe's Convention on Biomedicine and Human Rights, because critics felt that the ethical standards recommended were below national ones. Unlike most EU member states, Germany recently voted in favor of a UN Declaration prohibiting "human cloning," which is deliberately unspecific as to whether, as critics fear, it also applies to therapeutic cloning (see Abels in this book). In short, in the case of ART Germany has tried to extend its own restrictive standards to the international level, while supra- and international norms are negligible for explaining the restrictive national ART policies.

## CONCLUSION

Germany followed a restrictive regulatory path in both fields. There are parallels but also interesting differences regarding the explanatory variables (see table 7.3). The two fields differ clearly with respect to the international regime hypothesis. Europeanization is a key factor in explaining the GMO case, because it influences domestic opportunity structures, and compliance with the relatively strict EU policies is required. Nevertheless, Germany could have chosen a less restrictive implementation of EU policies on

**Table 7.3.  Explanatory Factors**

|                                                                        | ART | GMO |
|------------------------------------------------------------------------|-----|-----|
| Policy Network                                                         |     |     |
| H1: Policy-community of targets                                        | NO  | NO  |
| H2: Broad mobilization by critical interest groups in issue network    | YES | YES |
|                                                                        |     |     |
| Country Pattern                                                       |     |     |
| H3: Christian Democrats/Green Party in government / key player         | YES | YES |
| H4: Critical groups pushing on national and federal level              | YES | NO  |
| H5: Ministries in charge not sharing interests of target groups        | YES | YES |
|                                                                        |     |     |
| Internationalization                                                  |     |     |
| H6: Europeanization/European policies successfully used by critical interest groups | NO  | YES |
| H7: Policy transfer through competition                                | NO  | NO  |

the national level. The stringent EU policies strengthened the position of interest groups critical of biotechnology on the national level, hence contributing to restrictive German policies. In the field of ART, however, Europeanization and international norms are marginal. International competition is limited in both fields, yet favored the authorization of embryonic stem cell research in Germany. In the GMO field, German companies are latecomers often now focusing on the so-called second generation of GMOs. We can conclude that international multilevel governance matters only in one field, but that similar sector-specific variables interacting with institutional characteristics of the German political system explain the similarly restrictive policies in both fields.

There are, first of all, strong parallels with respect to the network variables, self-regulation, public mobilization and framing. They are due to the coupling of the issues of ART and GMOs during the 1980s in a "programmatic" debate about biotechnology and its potential impacts on society. The 1980s were characterized by the rise of the new left and the environmental movement. Strong public mobilization against biotechnology in general, in the case of ART supported by the churches and religious interest groups as well as feminists, contributed significantly to restrictive outcomes. In addition, Germany's past also explains the particularly strong fear of possible eugenic uses of biotechnology. At the same time, self-regulation by target groups did not prevent a strong state intervention, but imposed supplementary restrictions in both fields, in contradiction to the proposed theoretical hypothesis (see chapter 1).

Different decisional arenas were important in the two fields as a consequence of Europeanization. In the GMO case, the legislature played a comparatively less important role. Furthermore, the political party constellation differed across the two policy fields. Party positions were more homogenous in the field of ART, where government and opposition, right and left were advocating restrictive policies or even total prohibitions of IVF. In the GMO case, party differences were more important: while all parties advocated some form of regulation, they differed on the means and degree of restrictiveness. The fact that the Christian Democrats were in power at the time of the adoption of the EschG contributes to explaining the restrictive ART policies; the policies, however, would not have been less restrictive under a left government. In the field of GMOs, the Green Party as junior coalition partner of the SPD contributed considerably to a restrictive implementation of EU norms. Hence, these findings support the hypotheses about the influence of the governmental parties (CDU/CSU and Greens) formulated in the theoretical framework.

Furthermore, the fact that the ministries responsible for some of the important legislative projects did not necessarily share the interests of the target groups (in the case of ART, the Ministry of Justice elaborated the ESchG bill), which weakened the position of economic, research and medical interest groups. In the GMO case, the newly established Federal Ministry of Consumer Protection took over GMO policy, advocating for a change in agricultural policy (*Agrarwende*) focusing on consumer and environmentalists as the final beneficiaries, thus opposing the influence of the target groups (e.g., producers, researchers). These results confirm the theoretical assumptions in this respect.

Finally, the multiple points of access and veto characteristic of the German political system foster compromises. In the case of the EschG, the *Länder* played a more important role, with a clear majority pushing for restrictive policies; in terms of stem cell research, the *Länder*, with the exception of Northrine-Westphalia under a SPD government until May 2005, opted for a restrictive implementation of, and opposed liberalization of, the existing StZG. Yet, in the GMO case, positions differed along party lines: the conservative and liberal parties focused on the target groups advocating less restrictive legislation while the Green Party in particular, but also the SPD, were more responsive to consumer and environmental concerns. There is now, however, more disagreement within parties, namely in the SPD, on future policies, given the increasing popularity of framing biotechnology as a key technology for future economic success.

Despite the restrictive design in both fields, there is an interesting difference regarding the actual market (i.e., the availability of GMOs and ART). The GMO policy design is actually accompanied by a restrictive implementation: the number of field experiments is low and there are hardly any GMOs on the market; moreover, Germany has invoked the safeguard clause and established national bans. In contrast, restrictive policies do not prevent the high availability of ART services (see Nygren and Andersen 2002) at least for the groups granted legal access and for permitted techniques. The answer to this puzzle is the beneficiaries: although there is high collective pressure from consumer groups and environmentalists for restrictive implementation in the GMO field, there is high individual demand of patients for ART services. The question of the availability of GM food and the effective access to and use of ART services would be another interesting object of comparison, where not only legal norms, but consumer behavior and the capacity of national health care systems in delivering ART treatment would also have to be considered.

## NOTES

We would like to thank Celina Ramjoué, who contributed to the data collection for the ART field.

1. In 1994 the German Basic Law was amended to include ART as one of the domains in which the federal government has the prerogative to legislate. A comprehensive federal law on ART, however, has not yet been adopted.

2. The guidelines of the Bundesärtzekammer (German Medical Association) were subsequently updated in 1988, 1991 and 1998 in order to take technological progress and new legislation into account.

3. About five hundred hectares in Germany; cf. www.biotech-info.net/Germany_releases_ban.html; (March 22, 2005).

4. This result is theoretically supported by such theorists of political communication as Friedhelm Neidhardt or Hans Martin Kepplinger.

5. There is certainly a market for genetic testing kits, laboratory equipment necessary, etc., but not for IVF service, counseling and testing service, which is, at best, limited to paternity tests.

## REFERENCES

Bahnsen, U., and U. Willmann. "Stammzellforscher sehen schwarz." *Zeit*, May 25, 2005: 29–30.

Bandelow, N. C. 1997. Ausweitung politischer Strategien im Mehrebenensystem: Schutz vor Risiken der Gentechnologie als Aushandlungsmaterie zwischen Bundesländern, Bund und EU. In *Politik und Biotechnologie. Die Zumutung der Zukunft*. Ed. R. Martinsen. Pp. 153–68. Baden-Baden: Nomos.

Behrens, M., S. Meyer-Stumborg, and G. Simonis. 1997. *GenFood. Einführung und Verbreitung, Konflikte und Gestaltungsmöglichkeiten*. Berlin: Edition Sigma.

Benda Report. 1985. In vitro Fertilisation, Genomanalyse und Gentherapie. Bericht der gemeinsamen Arbeitsgruppe des Bundesministers für Forschung und Technologie und des Bundesministers der Justiz. Hrsg. vom Bundesministerium für Forschung und Technologie. München: Schweitzer.

Benz, A. 1999. Der deutsche Föderalismus. In *50 Jahre Bundesrepublik Deutschland. Rahmenbedingungen–Entwicklung-Perspektiven*. Ed. T. Ellwein and E. Holtmann. Pp. 135–53. Opladen: Westdeutscher Verlag.

Betta, M. 1995. *Embryonenforschung und Familie. Zur Reproduktion in Grossbritannien, Italien und der Bundesrepublik*. Frankfurt/M.: Peter Lang.

Epp, A. 2003. *Law in Conflict: The Regulation of Genetically Modified Food in Germany and the United States*. 2003, http://bieson.ub.uni-bielefeld.de/volltexte/2005/628/pdf/EppDiss.pdf (March 15, 2005).

Gaskell, G., N. Allum, and S. Stares. 2003. *Europeans and Biotechnology in 2002: Eurobarometer 58.0*. 2nd ed. (March 21). London.

Geyer, C. (ed.). 2001. *Biopolitik. Die Positionen*. Frankfurt/M.: Suhrkamp.

Gill, B. 2003. *Streitfall Natur. Weltbilder in Technik- und Umweltkonflikten*. Opladen, Wiesbaden: Westdeutscher Verlag.

Gottweis, H. 2002. Stem cell policies in the United States and Germany: Between bioethics and regulation, *Policy Studies Journal* 30 (4): 444–69.

Graumann, S. (ed.). 2001. *Die Genkontroverse*. Freiburg i. Br.: Herder.

———. 2002. *Situation der Medienberichterstattung zu den aktuellen Entwicklungen in der Biomedizin und ihren ethischen Fragen*. Gutachten für die AG "Bioethik und Wissenschaftskommunikation" am Max-Delbrück-Centrum für Molekulare Medizin Berlin, www.bioethik-diskurs.de/documents/wissensdatenbank/gutachten/Download-Dokumente/Graumann-Gutachten (July 5, 2005).

Hampel, J., G. Ruhrmann, M. Kohring, and A. Görke. 1998. Germany. In *Biotechnology in the Public Sphere: A European Sourcebook*. Eds. J. Durant, M. W. Bauer, and G. Gaskell. Pp. 63–76. London: Science Museum.

Hampel, J., and O. Renn (eds.). 2001. *Gentechnik in der Öffentlichkeit: Wahrnehmung und Bewertung einer umstrittenen Technologie*. Frankfurt/M., New York: Campus.

James, C. 2004. Preview: Global Status of Commercialized Biotech/GM Crops: 2004. ISAAA Briefs No. 32: Ithaca, NY: ISAAA.

Kepplinger, H. M. 1995. Die Gentechnik in der Medienberichterstattung. In *Gentechnologie in Deutschland*. Eds. Wolfgang Barz, Bernd Brinkmann, and Hans-Jürgen Ewers. Pp. 195–213. Münster: Lit-Verlag.

Kingdon, J. W. 2003. *Agendas, Alternatives and Public Policies*. 2nd Ed. New York: Longman.

Lassen, J., K. Madsen, and P. Sandøe. 2002. Ethics and Genetic Engineering: Lessons to Be Learned. *Bioprocess and Biosystems Engineering* 24: 263–71.

Lijphart, A. 1999. *Patterns of Democracy: Government Forms and Performance in Thirty-Six Countries*. New Haven: Yale University Press.

Niclauss, K. 1999. Bestätigung der Kanzlerdemokratie? Kanzler und Regierungen zwischen Verfassung und politischen Konventionen. *Aus Politik und Zeitgeschichte* B20: 27–38.

Nygren, K. G., and A. N. Andersen. 2002. Assisted Reproductive Technology in Europe, 1999: Results Generated from European Registers by ESHRE. *Human Reproduction* 17 (12): 3260–74.

Pollack, M. 1997. Representing Diffuse Interests in EC Policy-Making. *Journal of European Public Policy* 4 (4): 572–90.

Skogstad, G. 2003. Legitimacy and/or Policy Effectiveness? Network Governance and GMO Regulation in the European Union. *Journal of European Public Policy* 10 (3): 321–38.

Waldkirch, B. 2004. *Der Gesetzgeber und die Gentechnik. Das Spannungsverhältnis von Interessen, Sach- und Zeitdruck*. Wiesbaden: Verlag für Sozialwissenschaften.

Wilde, G. 2001. Das Geschlecht des Rechtsstaats. Herrschaftsstrukturen und Grundrechtspolitik in der deutschen Verfassungstradition. Frankfurt/M., New York: Campus.

# 8

# ACCOMMODATION, BUREAUCRATIC POLITICS, AND SUPRANATIONAL LEVIATHAN: ART AND GMO POLICY-MAKING IN THE NETHERLANDS

*Arco Timmermans (University of Twente)*

The Netherlands is the world's third largest exporter of agro and food products, following the United States and France (Oosterwijk 2003: 351). This fact is due to a broad range of food-related economic activities, including the transit trade, but it is also the result of a specialization in biotechnology. In contrast to the international orientation of the Dutch agro-food industry, the medical community in charge of assisted reproductive technology is much more limited in its geographic scope. Nevertheless, the medical community has much greater leeway in using assisted reproductive technology in medical practice than the agro-food industry's ability to introduce GMO food products into the market.

The Netherlands is known as an open country with a political system that does have its peculiarities. The Dutch parliamentary system has been referred to as a "centralized consensus democracy" (Lijphart 1999; Timmermans 2001). In this type of democracy, institutional arrangements for policy-making are based on principles of proportional representation and power sharing. The legislature contains multiple political parties that reach across socioeconomic and religious cleavages; governments are always coalitions. The building and maintaining of coalitions involves considerable political costs, particularly when dealing with moral issues, such as abortion, euthanasia and assisted human reproduction. Frequently, policy decisions in these fields are sacrificed for the continuation of the government. It is not unusual for parties to commit themselves to political deals that guarantee nondecisions before they join office (Timmermans 2003).

In the case of the Netherlands, political representation and the involvement of interest groups are organized according to formal and functional rules, that is, rules specifying access, competencies, and resource allocation, for both the actors responsible for political decision-making and the various stakeholders in the related policy field. In the fields of ART and GM food, the stakeholders include scientific researchers, members of industry, end-users, and representatives of government ministries. All these actors have knowledge, money or competencies as assets. Between the two policy fields, there are differences in the size of the stakes at play and, as a consequence, in the relative significance of actors. In this chapter, we analyze this variation and show its implications for policy-making.

The chapter begins with an account of the development of ART and GMO policies since these biotechnologies arrived on the political agenda in the 1980s. We then present possible explanations for the policy processes and their results, as well as for the variation between the two fields. We then consider the extent to which present policy packages and their differences can be explained by the policy networks that have formed around these issues. We also analyze the impact of the properties specific to the national system within the context of Parliament, the government, and at the level of interdepartmental relationships. And finally, we examine the role of international regimes and multilevel governance on these two fields.

## ART POLICY: BEYOND PROFESSIONAL AUTONOMY

Since the mid-1980s, government intervention in ART has been the subject of debate, although primarily limited to a select group of actors (Timmermans 2004). In the 1980s, the framing of the problem for ART focused on infertility, which was seen as a medical problem for which the emerging technologies offered a solution. The medical community was not the only group to hold this view, but it was also supported by FREYA, the Dutch Association for In Vitro Fertilization, an organization representing client interests. The emerging technologies were, initially, "in search of a public" (Kirejczyk 1996: 88); once in use, it was believed that these technologies required state planning and control. Past experiences with the abortion policy debates of the 1970s and early 1980s deterred political parties from engaging in a debate on the topic and led them to circumvent policy choices by focusing mainly on procedural aspects.[1] Thus the government delegated to the Ministry of Health the task of putting together a compulsory licensing system for IVF, based on a report of the Health Council. In 1989, the In Vitro Fertilization Planning Decree (Planningsbesluit in vitro fertilisatie[2]) was published. This decree allowed forty-five hundred IVF treatments per year, allocated among eleven licensed hospitals. Arrangements for the subsidization of IVF were halfheartedly made, insofar as IVF treatments would be financed by the Health Insurance Council, but would not be formally incorporated into the Health Insurance Fund, as this remained a politically controversial issue.

In the early 1990s, the dominant image of a planning and control approach to the policy design fell into disfavor, because the containment of medical practices and research was less effective than expected. Concerns arose in light of activities involving aspects of embryo research in medical centers that were currently unregulated. Furthermore, the interests of children born after medical intervention became more prominent among doctors, social groups, and politicians. These factors contributed to the perception that greater control was required (Kirejczyk 1999: 896; Van der Bruggen 1999: 16–17). Despite political commitments being made to initiate legislation, agreement on the proposed bills was difficult to achieve. In the mid-1990s, the traditionally powerful Christian Democrats were no longer in government, which led to existing legislative projects on ART being critically reviewed and a number of new bills being prepared.[3] This new government was the first to consider research on embryos used for purposes other than reproduction. The medical associations also took the initiative by producing self-regulation, which was often in line with the reports issued by the Health Council. In 1998, the rapid increase in the proportion of clients applying for IVF treatment for reasons other than blocked fallopian tubes, which was the existing condition for treatment, prompted the Association of Obstetricians and Gynecologists to specify a new set of conditions for IVF treatments.[4] Also in 1998, the association presented a set of internal guidelines for surrogacy, in particular the containment of the supply and demand of egg cells.[5] This self-regulation was meant to prevent uncontrolled practices, but it also demonstrated that the medical profession remained largely autonomous.

The government subsequently updated the regulatory framework in 1998.[6] Gender selection was prohibited, except when medical grounds could show that it could prevent a gender-related genetic disease in the child.[7] Other changes in regulation included the permission of surrogate motherhood on conditions established by the NVOG, and a relaxation in the rules for subsidizing IVF by the Health Insurance Council, which increased the number of IVF centers covered by the subsidy.[8] The Public Health Insurance Fund covered three IVF treatments per patient; however, governmental budget cuts in late 2003 ended the subsidy for the first IVF treatment, for which patients thus needed to obtain additional insurance. In June 2000, single women or lesbian couples were no longer excluded from medical treatment. The IVF Planning Decree was amended in December 2000 to allow a limited number of academic medical centers to conduct research on MESA;[9] however, the moratorium on practicing this technique remained in place.

In June 2002, the important Embryo Act, which had been prepared by the government that still excluded the Christian Democratic Party (CDA), was passed by Parliament, despite the CDA voting against it in both Houses. Under the Embryo Act, the scope of legislation on assisted reproductive technologies increased significantly.[10] The act included rules on the control of and property aspects of gametes and embryos, on the conditions for consent by adults providing gametes or embryos, on embryo research for purposes of pregnancy or other purposes, on the use of fetal tissue, and on the prohibition of gender selection for nonmedical grounds; it also states that a central commission should monitor the implementation of the act, supervise medical centers doing embryo research, and annually report to the minister of health.

The scope of the government's ART policy is now much broader than it was before the turn of the century (see table 8.1). The updated IVF Planning Decree (1998, amended in 2000) and the Embryo Act of 2002 constitute a policy package that contains permissive as well as some restrictive elements, but which is intermediate overall. The goals contained in this set of regulations are both procedural and substantive: guaranteeing high quality and efficiency in IVF medical practices, while also safeguarding self-determination and human dignity. Although the first goal lends support to what was already socially and politically accepted, the second goal concerns the boundaries of what is considered acceptable.

The IVF Planning Decree and the Embryo Act contain procedural and substantive constraints on the freedom of maneuver of the medical community, applying both to practitioners and researchers. The Embryo Act led to a change in the Medical Research Involving Human Subjects Act (*Wet medisch-wetenschappelijk onderzoek met mensen*, WMO), requiring formal *ex ante* review of embryo research by the Central Committee on Research Involving Human Subjects (*Commissie mensgebonden onderzoek*, CCMO). Constraints on IVF and embryo research, however, did not all originate with the government; constraints were adopted in part from self-regulation

**Table 8.1.   Current ART Policy Design**

|  | Research | Practice of fertility treatments | Measures aimed at patients |
|---|---|---|---|
| Policy Design | Prohibition of: creation of human embryos for research hybrid and chimera building genetic modification of embryos; Conditions: research on left-over embryos from IVF only with permit from Central Committee on Research involving Human Subjects; written consent from cell donor required | Treatments only in licensed centers; new fertility techniques can be applied only after Medical Technology Assessment | Prohibition of commercial surrogacy; collective health insurance does not cover first IVF cycle; in cases of AIDs, child has right of donor information |
| Type | Restrictive | Permissive | Intermediate |

produced by the Dutch Association of Obstetricians and Gynecologists. Centers may install Medical Ethics Review Committees, which must be recognized by the Central Committee on Research Involving Human Subjects (CCMO). The list of conditions for IVF treatment compiled by the NVOG has expanded gradually. Practitioners enjoy considerable autonomy in treating a more diverse clientele than the married heterosexual couples for which the technology was initially meant. The Planning Decree also contains clear guidelines, such as the rule that a surrogate mother should have first given birth to at least one healthy child herself. Commercial organizations seeking access to reproductive technologies and their results face very strict national conditions and often do not have access at all. Given that IVF treatments are subsidized by the Public Health Insurance Fund, individual clients have financial access to IVF, though a recent government decision to exempt the first IVF treatment from this public subsidy limits this financial access point. Another financial constraint on access to IVF is that private insurance companies sometimes exclude former Public Health Insurance customers when undergoing IVF treatment.

The Embryo Act imposes additional substantive constraints on the autonomy of the medical community, the main target group of this legislation. In particular, the Embryo Act prohibits the creation of embryos specifically for research, except in cases where research is intended to increase knowledge of infertility, assisted reproductive technologies, genetic diseases or transplantations. In March 2004, the Central Committee on Research Involving Human Subjects (CCMO) gave the first permit for stem cell research using human embryos under the current legal constraints. Also prohibited are such practices as in vitro development of embryos beyond fourteen days, cloning, genetic manipulation, hybrid and chimera building, and implantation of human embryos in animals or animal embryos in humans. The prohibition against the creation of embryos specifically for research however was meant to be temporary; the Embryo Act states explicitly that the government will propose a decision to drop the prohibition within five years. This political commitment to a restrictive interpretation of the law was meant to last until the parliamentary elections in 2007. However, the Christian Democrats returned to office in November 2002, after an eight-year absence, and the new government explicitly abandoned this last commitment; the subsequent government, formed in July 2003, confirmed this position.

## GM FOOD POLICY: THE LIMITS OF INCENTIVES FOR TECHNOLOGY

When national government policy on modern biotechnology began, the intention was not to restrict but to promote scientific developments and innovations benefiting the national economy. A strategy of codification rather than regulatory intervention was pursued in combination with the financial promotion by the national government of biotechnology as an emerging field.

Before the late 1970s, GMOs did not play an important part in government policy. The regulatory and legislative framework in place in the 1970s was based on a tradition in which organized interests of producers in the agro-food sector were well represented, and jurisdictions of government ministries and their agencies were clearly defined. Thus the Food and Drugs Act (*Warenwet*), originally from 1919, provided the legal basis for the Food and Drugs Agency, presently called the Food and Drugs Authority, the government agency in charge of food quality control. The Food and Drugs Act is still prominent, but it has become surrounded by GMO-specific regulations produced in arenas of multilevel governance. Public spending to support research and development was organized in a special program, the Innovation Oriented Biotechnology Program, which was launched in 1981 and ran until 1987 (Oosterwijk 2003: 418).

National parliamentary discussions on genetic modifications started in the late 1970s, when the opportunities and risks of innovation in the emerging technologies were beginning to be acknowledged. In 1981, the first regulation on genetic modifications was adopted by Parliament, and was incorporated into the Nuisance Act (*Hinderwet*). This regulation dealt with contained use (i.e., experiments taking place in research institutes). This regulation however was not meant to constrain producers but to fix a set of procedures in accordance with an emerging industry practice, as the government did through the IVF Planning Decree in the domain of ART in 1989. The important difference with ART however is that in the field of GMOs, industry self-regulation was absent. Competitive forces in this domain hindered such collective action by private industry.

As modern biotechnology developed and the discussion of field experiments began in the 1980s, regulations on field releases were adopted in 1990, which subsequently became known as the important GMO Decree.[11] This decree was based on the Hazardous Substances Act (*Wet milieugevaarlijke stoffen*), but it was not a formal law. This fact was signaled by the Council of State, the central advisory body of the government considering legal issues, which argued that the matter of regulating GMOs was so important that formal and more comprehensive legislation was required (Advies Raad van State, July 12, 1989, nr. W08.89.0122). The government, however, refrained from drafting such legislation, as it anticipated problems of viability and implementation, a view that was supported with little debate in the 1991 Parliament (Tweede Kamer vergaderjaar 1991–1992, 22 300, nr. 11). Since that time, the key pieces of national regulation for GM food are the GMO Decree and the Novel Food Decree, which conform to European Directives 98/81 and 2001/18. Other relevant elements of the emerging national regulatory regime are Article 8 of the Environmental Protection Act, which lists the sites for experimentation that require a license. In most cases, local administrations are attributed responsibility for licensing (Staatsblad, 1994: 80). A Decree on Accommodations and Licenses (*Inrichtingen- en vergunningsbesluit*) enacted on January 5, 1993, further specified the implementation of Article 8 of the Environmental Protection Act, and mentioned GMOs as a special category for which additional licensing conditions existed (Staatsblad 1993: 50). In 1995, the ministries involved in policy implementation and monitoring evaluated this regulation and concluded that the rules could be relaxed wherever this was warranted by the accumulated experiences (Ministerie van Algemene Zaken et al. 1995). In the late 1990s, the government stated that non-GM production chains should be supported "whenever possible," and this intention to promote a coexistence of GM food and non-GM food has been upheld since then (TK 1997–1998, 25 126, nr. 5).

At the same time, the financial incentives for biotechnology provided by the national government were weakening. In the 1990s, public spending for biotechnology research declined, with the rate of investment in the Netherlands dropping to a low, at times even the lowest, position in Europe (Enzing 2000: 109–10; Van Beuzekom 2001). This downward trend triggered criticisms from economic analysts, including one who concluded that "apparently the Dutch do not have biotechnology in their genes" (Financieel Economisch Magazine 2000). This decreasing support for public research and development was due in part to a Dutch feature of the allocation of public funds, in which a proportional distribution was made between different sectors of the economy, including biotechnology, and also within the field of biotechnology (Enzing 2000: 116). Though the Dutch agro-food sector was well established in the international market, industry expressed its concerns. The Ministry of Economic Affairs signaled a decline in the number of new commercial biotechnology enterprises (Ministerie van Economische Zaken et al. 2000).

At the turn of the century, the government reconsidered its efforts in public spending on biotechnology and, at the same time, the public debate, which had begun in the 1980s, increased in scale and scope. The acknowledged decline in research and development activities in biotechnology led the Ministry of Economic Affairs to launch a Life Sciences Action Plan for 2000–2004. Another effect was the installation of a Temporary Advisory Committee for Knowledge Infrastructure Genomics (*Tijdelijke Adviescommissie Kennisinfrastructuur Genomics*), which in 2001 launched the National Genomics Initiative, a program targeted specifically at research institutions, with a budget of $\varepsilon$ 186 millions. Further plans for major financial injections were presented in 2002, and food and biotechnology were mentioned as key themes (WRR 2003: 254).

These initiatives for financial support were a welcome push towards technology that had been long awaited by research institutes and industries facing international competition. But at the same time, stricter constraints on the commercial use of GMOs in food products were placed on the agenda. GM food officially became an issue of public debate when the government presented a multidepartmental white paper on biotechnology (Ministerie van Economische Zaken c.s., Integrale Nota Biotechnologie TK 2000–2001, 27 428, nr. 2). This white paper was presented as a strategic document, but also represented a compromise between six government departments with strongly conflicting interests. Thus the white paper was an attempt to accommodate jurisdictional struggles and balance environmental and economic criteria for policy-making. The increasing emphasis on the precautionary principle, however, was an important achievement of the Ministry of Environmental Affairs, which was the most critical of GM food. More generally, ethical values and social concerns became more prominent in the public and political debate on biotechnology, in which GMOs were a focal point.

The policy package on GM food in place by the end of 2004 built on the aforementioned developments in terms of financial support, attempts at social dialogue, and on the emerging regulatory framework within the European Union. As a result, the policy has become more restrictive, despite the recent financial injections in biotechnology research. On this last point, the government continues its attempts to accommodate enthusiasm for and skepticism against technology by financing, through the Netherlands Genomics Initiative, a research program on the nontechnical aspects of genomics, to which GM food relates closely. More generally, the national government is expected to take an intermediary position between industries and citizens, and between national actors and supranational institutions (WRR 2003: 209–10). This is why the Ministry of Agriculture, Nature and Food Quality promotes a policy of coexistence of GM crops and conventional and organic crops (Ministerie van Landbouw, Natuurbeheer en Voedselkwaliteit 2003). This policy reflects the intentions of the Balkenende II government composed of Christian Democrats (CDA), Liberals (VVD) and Liberal Democrats (D66) not to engage in extensive government regulation and promote self-regulation within the sector.

As we saw in the chapter by Gabriele Abels, the key pieces of European regulation in place by mid-2004 are the Directive 98/81/EC on contained use, the Directive 2001/18/EC on field and market release, and Ordinance COM (2001) 182 on traceability and labeling, which was adopted by the European Council of Ministers on July 22, 2003. In addition, a proposal for Ordinance COM (2002) 85 that regulates cross-boundary transport of GMOs is currently under scrutiny.

Directive 98/81 and Directive 2001/18 were both implemented by the national government. In September 2004, the GMO Decree on contained use and field release was updated in accordance with the European rules, thus tightening reporting and information requirements.[12] Risk assessment was sharpened, but the principle of proportionality, which also requires an assessment of the expected utility, was not mentioned. A revision of the Ministerial Order on GMOs (*Regeling GGO*)[13] was enforced in June 2004, essentially a national implementation of Directive 98/81. The Committee on Genetic Modification (COGEM) formulated the substantive and procedural rules for this ministerial order. The Decree on Accommodations and Licenses (*Inrichtingen- en vergunningenbesluit*) enacted in January 1993 gave conditions for licensing, which remain in use today. Ordinance 258/97/EG on Novel Foods was also implemented into the Novel Foods Decree (*Warenwetbesluit nieuwe voedingsmiddelen*), which gives conditions for products that are marketed as "gentech free."[14] Finally, the European Ordinance 1830/2003 on traceability took force on April 18, 2004.

The underlying principles for this policy package were derived in large part from the European regulatory framework. Since this European framework emerged in the 1990s, political actors in the Netherlands, including most prominently the national government, adopted the principles of the framework (i.e., at the level of symbolical politics). But despite the con-

**Table 8.2.   Current Dutch GMO Policy Design**

|  | Field trials | Commercialization and cultivation | Consumer and environment |
|---|---|---|---|
| Policy Design | EU deliberate release directive transposed into GMO Decree and GMO Regulation. GMO Office is licensing authority, decides case by case on advice COGEM; monitoring by Inspectorate for Environment; legal and financial sanctions | EU law for GM food and feed, and deliberate release; labeling requirements; monitoring by Environment Inspectorate and Food and Product Authority; financial and legal sanctions | EU law for GM food and feed, deliberate release, and traceability and labeling; monitoring by aforementioned agencies; consumer information by Food Center |
| Type | Intermediate | Intermediate | Intermediate |

verging effects of this European framework, the emphasis on such values as the precautionary principle and proportionality still vary between ministries and also in the national regulation. This may explain part of the variation in the degree of restrictiveness of GM food policy in the Netherlands, as table 8.2 shows. In the important GMO Decree the proportionality principle is seen to be insufficiently present, and the COGEM recently advocated a revision of this decree (COGEM 2003: 35).

The overall profile of regulation on GM food is intermediate. The more limiting rules focus on the market release of GM foods, to which the recent regulatory requirements on traceability and labeling have contributed. These rules set conditions for research and development on "green" biotechnology to the extent that such research efforts are driven by commercial goals for the European food market. But public and private research centers have broader and longer-term agendas, and existing uncertainties encourage rather than discourage further research investments. At the same time, attempts at dialogue among public and private stakeholders will likely continue in the coming years.

## EXPLAINING VARIATION BETWEEN ART AND GM FOOD POLICY-MAKING

Dutch ART policy shows a tendency to restrict genetic research, whereas GM food policy is becoming more restrictive on conditions for market access. This recent variation in emphasis, in terms of regulating elements at different points in the production chain, whether from laboratory to end-users or from farm to fork, is rooted in the differences in the extent to which the supply and demand for technologies and their uses in these two domains was, traditionally, the dominant coordination mechanism. Against this background, policies in these high-technology fields have resulted from a set of actors engaged in such different arenas as sectoral self-regulation, economic competition, controversies over social and political values, and European regulation. The contexts of these arenas and actors differ with respect to ART and GMO policy; this section examines these differences and their impact on policy-making.

### Policy Networks: Keeping Windows of Opportunity Closed

We may ask to what extent the emerging policy and the resulting processes can be determined by the conditions set by the networks of actors within the policy subsystem. Such subsystems contain not only structural arrangements but may also be the immediate context in which images of problems and solutions are created. Policy networks may contain multiple types of actors, and be exclusive or inclusive (i.e., more or less open to the contributions of new players). Policy networks cut across the public and private divide. In the Dutch case, they often contain semipublic institutions that build on a long tradition of corporatism. These semipublic institutions exist in both of our policy fields, with the Public Health Insurance Council in charge of the Public Health Insurance Fund (ARTs) and the Commodity Boards (*Productschappen*) and the recently created Project Group Biotechnology (GM food) being good examples. In addition, expert advice organs also include spokespersons from a variety of institutional spheres, and often contain rules to guarantee the diversity of the composition of their members.

Initially, the policy networks that had emerged around ART and GMOs were both closed in terms of access. This may not be typical for such relatively new issues (which reached the political agenda in the mid-1980s), and the main reason for this was the historical institutionalization of relationships in the Dutch health sector and the agro-food sector. The public and semipublic institutions in these sectors were themselves "legacies of agenda access" (Baumgartner and Jones 1993). The "iron triangle" of agriculture was a long outstanding example of corporatism in the Netherlands, and the medical community had a strong tradition of professional autonomy. Thus, when confronted with the new issues of ART and GMOs, the sense of urgency for government regulation was itself a subject of accommodation among actors with stakes in these fields. This provides another part of the explanation for the emphasis on procedural regulation for ART and the initial reluctance to impose drastic restrictions on the food industry. Moreover, in the 1990s, the broader public in the Netherlands was relatively inclined to accept emerging applications of biotechnology and also showed confidence in biotechnology industries (Eurobarometer 2003). In other words, the relatively closed nature of the policy networks did not entail a serious problem of legitimacy in the eyes of the public. And thus far no policy entrepreneurs have managed to successfully expand the scope of debate and generate broad public attention.

In terms of ART, the number of players involved in policy-making has been limited, and the subsystem in which these players interact was relatively closed. To an extent, this was the result of the high degree of institutionalization of the broader domain of health policy in the Netherlands, where the organized medical profession enjoyed considerable autonomy and the research community was represented in expert bodies with a fixed position in the early stages of the policy process. In this small actor network, the role of the Ministry of Health was relatively limited; the network of actors involved in ART policy-making contained no clear power center (Kirejczyk et al. 2001). At the same time, the policy community was relatively inaccessible—patients, or clients, demanding medical treatment typically had much less access than organized professions. Furthermore, the Health Council was the primary body of expertise, and its recommendations on ART have been the main wellspring for policy decisions (Schroër 2001). The most recent regulation was the result of a shift to the political arenas when a window of opportunity presented itself after the 1994 parliamentary elections. But even this shift of institutional locus did not immediately lead to success—in the mid-1990s several attempts at producing legislation failed. Though these initiatives were never really vetoed from within the policy community, the legislative process was stalled by several rounds of consultation in which the Health Council was once again prominent (Gezondheidsraad 1997a; 1997b; 1999). The Embryo Act, however, explicitly prohibited, at least temporarily, the creation of human embryos for scientific research, an important point on which it deviated from the recommendations made by the Health Council, and which qualified as the political breakthrough of 2002.

With legislation on embryo research in place since that year, the set of further delicate issues regarding embryo research was transferred back to the policy subsystem. The policy monopoly built up since the 1980s was thus perpetuated, due not only to institutional structures but also because of the stability of the actors' socially constructed position. The image of the medical profession as "advantaged" (Schneider and Ingram 1997) continued to manifest itself in the apparent trust of the state in the profession's capacity for self-regulation, as well as within the extended national regulatory framework. As Klein (1993: 204) argues, conceding medical autonomy is a politically rational strategy to diffuse blame. Social groups in explicit opposition to ART, who had previously been portrayed as "deviants," continued to have difficulty in getting access to policy-making arenas or getting their voices heard in advisory bodies. The main client organization, FREYA, was mostly consulted on an ad hoc basis. In the subsystem, there was no acknowledged group of "contenders" making a common front against further advances in research and medical practice. These aspects of the policy network help to explain policy-making, the resulting intermediate policy regime and the closed policy subsystem.

The part of the biotechnology sector dealing with GM food involved multiple stakeholders, with a concentration of members from industry. Organizations for public research were decentralized, and jurisdictions were allocated among different government ministries, which was, in part, an institutional response to the emergence of new issues surrounding the technologies. Assouline and Joly (1999: 14–15) depict the Dutch biotechnology sector as concentrated and pluralistic. This characterization mirrors a gradual change from a solidly closed to a more open policy network. This can be attributed to the manifestation of departmental rivalries and the ongoing internationalization, which is beyond the scope of influence of players in the Dutch policy domain. Bernauer and Meins (2003: 647) argue that multilevel governance entails more venues for NGOs than centralized regulatory systems. In addition to this point, the structural features of the Dutch GM food policy sector also entail a *functional* decentralization. The multiple departments involved increased the number of venues for agenda access. The current policy package is much less the result of policy predetermination within the subsystem than it used to be. Moreover, the important Committee for Genetic Modification (COGEM) recently revamped its strategy and explicitly incorporated ethical and social aspects of biotechnology in its considerations and gave more prominence to public consultation in policy advising. With the integrated framework for *ex ante* evaluation of modern biotechnology, the COGEM has taken a position at the forefront of the European Union (COGEM 2003: 21). This development is also clearly driven by the acknowledged need to accommodate, from within the policy subsystem of GM food, the broader and increased public concerns about GMOs. A so-called Temporary Committee on Biotechnology and Food, chaired by the former vice-prime minister Jan Terlouw, organized a public debate and issued a report in January 2002 to the Minister of Agriculture; the implications of the report's main conclusions for policy-makers were that a more explicit and transparent framework for licensing was required (Eten en Genen, 2002). The national Parliament debated the issue and delegated the task of designing such a framework to the Minister of Environmental Affairs, who consulted the Committee on Genetic Modification (COGEM). The COGEM, the expert committee considering also ethical and social aspects of GMOs, presented a report in June 2003, and advised the government to employ an "integrated ethical and social frame for reviewing modern biotechnology" and to follow a stepwise and case-by-case approach to new technologies, based on the principle of proportionality in risks and benefits. Furthermore, the Ministry of Environmental Affairs was advised that its mandate extended beyond its departmental expertise to include ethical and social knowledge (COGEM 2003: 9–10). As noted, the COGEM not only advises the government on strategic policy issues, but is also consulted by the GMO Office in the process of licensing. This institutionalized position facilitates the practicing of principles presented by the COGEM in specific cases of licensing.

## Country Patterns: Accommodating Departmental and Party Political Conflict

The impact of political conditions on ART and GM food policy is large, but quite different between the two fields. As noted in chapter 1, the Dutch political system is a centralized consensus democracy. Subnational levels of government do exist, but were not relevant in ART and GMO policy-making.

In both fields, technological developments and the issues involved reached the public and political agenda in the 1980s. Before then, regulatory constraints on modern biotechnology were not salient, and ART was still in the early stages of technological development and fell within the domain of the autonomous medical community. Beyond this temporal similarity in agenda setting, the national routes to the current policy packages were rather diverse.

The increasing salience of ART in medical practice since the mid-1980s did not lead to a quick government response. The government took recourse to procedural regulation for most of the 1980s and 1990s. This apparent strategy can be seen as the result of the divisiveness of ART among the political parties in the national policy-making arenas. Value controversy among political parties and the institutional rules of the game motivated the governments in office since the late 1980s to follow strategies of depoliticization. The most important parties in the national Parliament were those supporting the majority government, and, until 1994, the Christian Democrats (CDA) played a pivotal role in government coalitions. The Christian Democrats were reluctant to go into the details of substantive regulation, not because they favored a policy of laissez-faire, but because they anticipated the difficulty of reaching agreement with the other key parties, the Social Democrats (PvdA) and the Liberals (VVD). As with other issues of morality policy, the three main parties, with the Christian Democrats in a strategic medium position, avoided value controversy. On such matters, parties in and outside government coalitions are notoriously incapable of making collective decisions (Timmermans 2003). The use of expert advice was the primary mechanism of depoliticization. Until the mid-1990s, an implicit consensus existed such that the government codified existing medical practices and relied on the principles and self-regulatory capacities of the medical community. Reports issued by the influential Health Council were used to substantiate this course of action. The codification was given the form of a compulsory licensing system in 1989, the IVF Planning Decree, which was revised several times since then.

But this implicit type of coalition politics was hard to sustain as technological developments continued and social groups began to put pressure on the medical community and the political parties to either increase the possibilities of access to medical treatment, or to put constraints on the scope of medical research activities. A political window of opportunity opened when in 1994 the Christian Democrats were defeated in the parliamentary elections and a government coalition of secular parties was formed. This coalition put legislation on embryo research on the agenda. Over time, the political beliefs on this medical research field had converged, which facilitated agreement on the Embryo Act that was adopted in September 2002, with the opposing CDA losing the parliamentary vote. Although this legislation was also based on expert advice, the Health Council played less of a role in depoliticizing the debate than before. This decreased reliance on the scientific advice of the Health Council fits a changing pattern of expert advice, but in terms of ART it was an exception (Bal et al. 2002).

Thus, most of the latent controversy on ART was about values and for some time this controversy was prevented from manifesting itself. Political confrontations were accommodated in the political arenas by effectively delegating substantive regulation to the medical community and by making use of expert authority. This mechanism, which is part of the broader politics of accommodation, was used less intensely between 1994 and 2002, and this resulted in important substantive legislation on embryo research. This legislation resulted from the political window of opportunity, but also from a sense of urgency motivated by international scientific advances in stem cell research. The return of the CDA to government in November 2002 did not directly affect this legislation itself, but will likely reduce the leeway of interpretation for any party in office together with the Christian Democrats.

If interparty conflict was most important in the policy process on ART, conflict in GM food policy-making was primarily interdepartmental. GM food policy-making typically involved bureaucratic politics rather than coalition politics. GM food issues and biotechnology generally speaking were only occasionally discussed in the political arenas. For example, in early 2002, a temporary parliamentary committee discussed the government white paper on biotechnology, which itself was a multidepartmental compromise. In the subsequent parliamentary debate a number of motions were adopted, mostly unanimously, which led to modest changes in government policy (Handelingen Tweede Kamer 2001–2002, 27428, nr. 46). This apparent consensus, achieved at low political costs, illustrates the uncontroversial character of the topic among the political parties.

GM food was thus much more a subject of bureaucratic politics. As described above, GM food, a key theme in modern biotechnology, involves multiple departmental interests. The greater salience of the market, the position of large and often multinational industries and the much broader group of direct beneficiaries of food regulation (compared to ART regulation) increase departmental stakes. The Ministry of Economic Affairs, which creates financial stimuli for research and development, did not oppose, as such, the regulation of GM food, and the Ministry of Agriculture with its long tradition of agro-food policy also acknowledged the need for regulation on environmental and health grounds. But as the risks of modern biotechnology and its applications in the agro-food sector increasingly placed against the benefits, a struggle of issue ownership and jurisdictions with the Ministry of Environmental Affairs came to the surface.

One important reason for the government to give Environmental Affairs regulatory primacy on GM food was the clear tendency in European and international biotechnology policy to put more weight on the environment. Thus since the late 1980s the Ministry of Environmental Affairs acquired an important role in national GM food regulation and the implementation of ensuing European Directives and Ordinances. As we saw, the COGEM was created as an expert body to advise the minister on GMOs, and to assist the GMO Office to deal with permits. The other ministry that was important for the issues of market release and consumer information (rules on traceability and labeling) was the Ministry of Health, which assigned the Committee on Novel Foods an advisory role.

Nevertheless, the involved ministries still encountered a clash of interests. The prominence of the European Union as a supranational Leviathan changed the struggle from one over regulatory power to one over implementation. This may have attenuated but did not remove the tensions between the national government departments involved in this field.

A final comparative observation at this level of analysis is that the uncontroversial nature of GM food among the political parties seems to have facilitated the national debates organized by the government. In the Dutch representative system, the use of public opinion or other venues of citizen voices tends to be considered with skepticism if the parties in the political arenas have fundamental disagreements. This may be why ART was not only long kept discrete in party and coalition politics, but was also avoided as a theme of broad public debate. The more limited size of the group of beneficiaries of ART policy may have been another reason, but the containment of the debate was an element of the strategy of depoliticization in this field. In this sense, ART policy-making fits the traditional image of Dutch politics of accommodation within the representative model of democracy, with its multiple attempts at depoliticization. GM food policy-making fits this model to the extent that the governing coalition mostly delegated policy decisions to bodies of experts and to ministers as heads of departments in charge of specific elements of GM food policy. If the emphasis on appeasement in this context of centrifugal forces could entail policy inertia, this did not happen. In the case of ART policy, this was because a major electoral shift led to a change in the government coalition in which the Christian Democratic veto on substantive regulation of embryo research was overruled. In the case of GM food, national progress in policy-making was made because of (and in part forced by) international mobilization of opposition by NGOs and the flow of EU regulation that resulted from it.

## Internationalization: Bottom-Up versus Top-Down

As in most other European countries, globalization and supranational governance was and is highly salient for GM food, but much less so for ART. The central expectation regarding globalization and European integration is an international convergence of policy.

In terms of ART, national commitments to supranational and international policy intentions were and still are relatively weak. This is primarily because these international intentions themselves contain soft commitments and are in part symbolic, as is shown in more detail in the chapter by Abels. In this field, international policy convergence occurs, if at all, mostly through voluntary lesson-drawing done in a bottom-up manner by national governments, and in the Dutch case this has not really happened. The national health system, as it expanded with the growth of the welfare state, always had a clear internal orientation, with an emphasis on policy coordination among the responsible ministries, the medical profession, and corporatist institutions for insurance and service delivery. We already saw this in the discussion of the closed policy community in this field.

In contrast, globalization and international policy convergence are visible and important in GM food regulation, particularly in recent years, and this trend is likely to continue. Until the 1990s, objections to the European policy track were made from within the Netherlands, such as objections to the shift from product- to process-based regulation, as well as the opposition to the European moratorium on the market release of new GMOs in April 1998. Recently, the Netherlands was within a group of countries facing a European Court of Justice case for not implementing Directive 2000/18 in time. Nevertheless, European harmonization of GM food regulation is strong, and, as Bernauer and Meins (2003: 655) assert, the EU has tended to do this at levels of stringency supported by the countries that are most averse to GMOs. Though national implementation of EU regulation is not automatically smooth, the leeway for noncompliance is decreasing. In short, the clear difference in the degree of globalization and multilevel governance helps to account for differences in both the process and content between the two policy fields. This difference however is not specific to the Dutch case; it appears for the most part in all European chapters in this volume.

## CONCLUSION

Dutch policies on ART and GM food have emerged in one national political system, in the same period since the mid-1980s, they are both regulatory in type, and both became denser and more restrictive in recent years. Yet important differences have emerged from the analysis in this chapter. Although the overall policy packages in both fields approach an intermediate position on the permissive-restrictive dimension, the emphases in these packages differ widely. This is important to acknowledge in representing the present policies.

The clearest point of variation in policy content is that regulatory restrictions on ART pertain to fundamental research, whereas for GM food they constrain market access, although they are still not as restrictive as, for example, in Germany. This ostensible asymmetry in regulatory restrictiveness, however, relates to one point, which is the social and political limit of acceptance of genetic modification given the perception of its effects on human life. Thus, in this social construction, the limits placed on GM food are based on the extent to which GM food is obtainable through the European and national markets for consumption and on the information about products available to consumers, whereas for ART the limit is placed on genetic research for human reproduction. For this reason the regimes for contained or field experiments with GMOs are more permissive; and this is also the case for medical treatments that do not involve genetic research on embryos. The established practice of ART is typically the result of a dominant framing of the problem in terms of infertility, with IVF and related techniques presented as the solution. The demand for this solution is still increasing. Likewise, modern biotechnology experiments in the agro-food sector are still seen in terms of their economic benefits; however, economic stakeholders are changing market orientations, from food markets to markets for technologies for genetic modification—to which a different regulatory regime applies.

This sectoral variation has also become visible in the analysis of the policy process. The difference in salience of the European and international context makes the hypothesis on the impact of an international regime formulated in chapter 1 more relevant for GMO than for ART policy-making. In the case of GM food policy, national regulation increasingly came under pressure of internationally mobilized opposition from NGOs, with a resulting recent flow of more restrictive EU regulation. Table 8.3 summarizes the hypotheses and their empirical underpinnings in the Dutch case.

Characteristics of the national political and bureaucratic system influenced the policy process and its results. In the Dutch systemic arenas, accommodation of value conflict is imperative, and this puts an emphasis on political appeasement rather than on setting clear policy priorities. In the case of ART policy, a breakthrough was possible because a major electoral shift in the mid-1990s led to a change in the government coalition in which the Christian Democratic veto on regulation of embryo research was even-

**Table 8.3.   Explanatory Factors**

|  | ART | GMO |
|---|---|---|
| Policy Network |  |  |
| H1: Policy community of targets | YES | NO |
| H2: Broad mobilization by critical interest groups | NO | YES |
|  |  |  |
| Country Patterns |  |  |
| H3: Christian Democrats/Green Party in government / key player | NO / YES | NO |
| H4: Critical groups pushing on national and federated level | NO (N/A) | NO (N/A) |
| H5: Ministries in charge not sharing interests of target groups | YES | NO / YES |
|  |  |  |
| Internationalization |  |  |
| H6: Europeanization/European policies successfully used by critical interest groups | NO | YES |
| H7: Policy transfer through competition | NO | NO |

tually overruled. The prominence of the supranational "Leviathan" moderated the interdepartmental struggles in the agro-food sector over biotechnology, but the conflict of interest has not disappeared. This case of Dutch GM food policy with its diverse interests represented by different ministries does not allow for clear conclusions about the hypothesis on the influence of administrative structures formulated in chapter 1. In fact, the rival forces that fought at the interdepartmental level characterize national policy-making in this field. On this theoretical point, ART policy-making was less complex at the interdepartmental level, thus providing more unambiguous evidence in support of this hypothesis.

The time perspective in this analysis of the policy history of regulation in these two fields of modern biotechnology is limited to less than two decades. Nevertheless, a conclusion that can be drawn is that the two policy subsystems have a propensity to limit the degree of participation and to sustain themselves, thus supporting the continuation of a "policy monopoly." In both cases, an institutional legacy of corporatist structures with fixed rules of entry and competencies existed, and the policy community rooted in this institutional context remained stable for a long time. Such an institutionalized relationship of policy-makers with scientific and economic stakeholders involved mutual gains, because the organized profession and industrial players remained relatively unconstrained by restrictive policy interventions, and policy-makers could avoid the political transaction costs of making such regulatory constraints.

But another conclusion is that policy subsystem stability is not absolute, and significant changes from outside can occur. If in the ART policy field, a degree of self-regulation effectively shielded the medical profession from substantive government intervention until the recent adoption of the Embryo Act (hypothesis one in chapter 1), this was much less a possibility for GMOs. In the case of GM food, these changes were the international mobilization of bias against GM food—expanding the system of participation—and the subsequent flow of European regulation. Thus, the case of GMOs, with its international "community of resistance," provides more evidence for another hypothesis formulated in chapter 1, namely the one on mobilization of critical groups. In the case of ART, the dynamics of agenda setting were triggered by a major national electoral shift and the unprecedented coalition change that followed from it. This sheds some interesting light on the hypothesis in chapter 1 on the influence of the Christian Democratic party. In the Dutch case, the Embryo Act containing restrictive conditions for embryo research was prepared and adopted during an episode of absence of the Christian Democratic Party from the cabinet and indeed of a relative powerlessness of this party on this issue in the legislative arena. However, when the CDA regained power in November 2002, it revoked a statement in the Embryo Act to drop the prohibition of embryo creation for research after five years. This last move, then, provides some empirical underpinning of the hypothesis.

## NOTES

1. The Abortion Act was approved in 1981 and took force in 1984. For an account of abortion policy-making in the Netherlands, see Outshoorn (1986).

2. Staatscourant, July 31, 1989.

3. Intrekking Wijziging van de Wet inzake medische experimenten i.v.m. regels inzake handelingen met menselijke embryo's en geslachtscellen (Handelingen Tweede Kamer, 1994-1995: 23016, no. 7); Notitie regelgeving inzake enige handelingen en wetenschappelijk onderzoek met embryo's en foetussen (Handelingen Tweede Kamer, 1994-1995, March 16, 1995).

4. Richtlijn Indicaties voor IVF (NVOG, 1998).

5. Richtlijn Hoogtechnologisch draagmoederschap (NVOG, 1998).

6. Planningsbesluit in vitro fertilisatie (Staatscourant, 1998, no. 95).

7. Besluit verbod geslachtskeuze niet-medische redenen (Staatscourant, May 26, 1998).

8. Wijziging regeling subsidiëring Ziekenfondsraad in vitro fertilisatie (Staatscourant, 1998, no. 84).

9. Wijziging Planningsbesluit in vitro fertilisatie (Staatscourant, December 14, 2000).

10. In March 2002, the government also submitted a bill on xeno-transplantation, containing a general prohibition. The government, however, left open the possibility that the responsible minister might redefine xeno-transplantation, which would then likely limit the scope of the prohibition.

11. Besluit GGO, January 25, 1990.

12. Besluit GGO, Staatsblad, July 9, 2003. This was the thirteenth revision of the decree since it was first adopted in 1990. This decree is based on the Hazardous Substances Act.

13. Regeling GGO, Staatscourant, May 22, 2004.

14. Warenwetbesluit nieuwe voedingsmiddelen, Staatsblad, May 15, 1997, 205. This decree is based on the Food and Drugs Act (Warenwet).

## REFERENCES

Allansdottir, A. et al. 2002. *Innovation and Competitiveness in European Biotechnology*. Brussels: European Commission.

Assouline, G., and P. B. Joly. 1999. The biotechnology policy-making and research system in the different countries: convergences and specifics. In European Commission, Inventory of public biotechnology R&D programmes in Europe (vol. 1). EUR 18886/1: 8–21.

Bal, R., W. E. Bijker, and R. Hendriks. 2002. *Paradox van wetenschappelijk gezag. Over de matschappelijke invloed van adviezen van de Gezondheidsraad.* Den Haag: Gezondheidsraad.

Baumgartner, F., and B. Jones. 1993. *Agendas and Instability in American Politics.* Chicago: University of Chicago Press.

Bernauer, T., and E. Meins. 2003. Technological Revolution Meets Policy and the Market: Explaining Cross-National Differences in Agricultural Biotechnology Regulation. *European Journal of Political Research.* 42 (5): 643–83.

Committee for Genetic Modification (COGEM). 2003. *Naar een integraal ethisch-maatschappelijk toetsingskader voor moderne biotechnologie.* Bilthoven: CGM/030618-02.

Enzing, C. 2000. *Dossier Biotechnologie.* Delft: TNO Strategie, Technologie en Beleid.

Eten en Genen. 2002. Een publiek debat over biotechnologie en voedsel [Eating and Genes. A public debate on biotechnology and food.] The Hague: Report of the Temporary Committee on Biotechnology and Food. Chaired by Dr. J. C. Terlouw.

Eurobarometer 58.0. 2003. *Europeans and Biotechnology in 2002.* Report to the EC Directorate General for Research.

Gezondheidsraad. 1997a. *Het Planningsbesluit IVF.* Den Haag.

———. 1997b. *Onderzoek met embryonale stamcellen.* Den Haag.

———. 1999. *Klinisch genetisch onderzoek en erfelijkheidsadvisering. Nadere advisering voor een nieuwe planningsregeling.* Den Haag.

Kirejczyk, M. 1996. *Met Technologie Gezegend? Gender en de omstreden invoering van in vitro fertilisatie in de Nederlandse gezondheidszorg.* Utrecht: Jan van Arkel.

———. 1999. Parliamentary Cultures and Human Embryos: The Dutch and British Debates Compared. *Social Studies of Science* 29 (6): 889–912.

Kirejczyk, M., D. Van Berkel, and T. Swiertra. 2001. *New procreation: Farewell to the stork.* Den Haag: Rathenau Instituut.

Klein, R. 1993. National variations on international trends. In *The Changing Medical Profession: an international perspective.* Eds. F. Hafferty and J. B. McKinlay. Pp. 202–209. New York: Oxford University Press.

Lijphart, A. 1999. *Patterns of Democracy: Government Forms and Performance in Thirty-Six Countries.* New Haven: Yale University Press.

Ministerie van Algemene Zaken e.a. 1995. *Regelgeving inzake genetisch gemodificeerde organismen. Evaluatierapport,* Den Haag.

Ministerie van Economische Zaken e.a. 2000. *Integrale Nota Biotechnologie.* Den Haag. Tweede Kamer vergaderjaar 2000–2001, 27 428, nr.2.

Ministerie van Economische Zaken. 2000. *Actieplan Life Sciences 2000–2004*. Den Haag.

Ministerie van Landbouw, Natuurbeheer en Voedselkwaliteit. 2003, *Beleidslijn coëxistentie van gg-gewassen, conventionele en biologische gewassen*. Den Haag.

Oosterwijk, H. 2003. *Sectoral Variations in National Systems of Innovation*. Doctoral dissertation: University of Utrecht.

Outshoorn, J. V. 1986. *De politieke strijd rondom de abortuswetgeving in Nederland 1964–1984*. s-Gravenhage: Vuga.

Timmermans, A. 2004. The Netherlands: Conflict and consensus on ART policy. In *Comparative Biomedical Policy: Governing assisted reproductive technologies*. Eds. I. Bleiklie, M. Goggin, and C. Rothmayr. Pp. 155–73. London, Routledge.

———. 2003. *High Politics in the Low Countries*. Aldershot: Ashgate.

———. 2001. Arenas as Institutional Sites for Policymaking: Patterns and Effects in Comparative Perspective. *Journal of Comparative Policy Analysis* 3 (3): 311–37.

Van Beuzekom, B. 2001. *Biotechnology Statistics in OECD Member Countries: Compendium of Existing National Statistics*. OECD, Directorate for Science, Technology and Industry, DTSI/DOC 2001/6.

Van der Bruggen, K. 1999. Dolly and Polly in the polder: debating the Dutch debate on cloning. In Rathenau Instituut, *The public debate on cloning: international experiences*. Amsterdam, November 19, 1999.

Wetenschappelijke Raad voor het Regeringsbeleid (WRR). 2003. *Beslissen over Biotechnologie*. Den Haag: Sdu.

# 9

# CONFLICT AND CONSENSUS IN BELGIAN BIOPOLICIES: GMO CONTROVERSY VERSUS BIOMEDICAL SELF-REGULATION

Frédéric Varone (Université de Genève) and
Nathalie Schiffino (Université Catholique de Louvain)

**B**elgium is a small European country that one would perhaps not expect to be a pioneer in the biomedical and agro-food sectors. Belgian scientific researchers, university hospitals and biotech firms have, however, contributed significantly to the development of the life sciences and to the commercialization of various biotechnological innovations. The first Intra-Cytoplasmic Sperm Injection was performed at the Vrije Universiteit Brussels in 1992, and the first genetically modified plant was developed at the University of Gent in 1983. Given Belgium's innovation in the promotion of both *Assisted Reproductive Technologies (ART)* and *Genetically Modified Organisms (GMO)* one might expect a convergence in the biomedical and agro-food policy fields. The approach to the collective problem to be solved in both fields is identical, that is, avoiding the negative effects of new biotechnologies, while favoring their positive applications. The expected convergence does not come to pass as Belgian ART policy is much less interventionist than the Belgian GMO policy.

This chapter attempts to explain this unexpected difference between the *permissive* and consensual regulation of biomedical, or red, biotechnologies and the *intermediate* and conflictual regulation of agricultural, or green, biotechnologies in Belgium. We identify three main explanations for these diverging biopolicies.

First, we begin with an examination of the important role played by *policy networks*. In the field of biomedicine, the medical actors and few political entrepreneurs constituted a closed "policy community." The professional interests of physicians and researchers, who favored a permissive regulation, clearly dominated. In contrast, the GMO sector was characterized by an open "issue network" that was composed of scientists, private biotech firms, environmentalist groups, consumer associations, as well as political and administrative actors. This broad range of competing research, economic, environmental and social interests resulted in a regulation of intermediate restrictiveness.

Second, three *specific characteristics of the Belgian polity*—such as the key role of political parties, the federalist structure and the (non) existence of an administrative agency—partly explain the differences between ART and GMO policies. Regarding political parties, the Christian Democratic party (coalition leader until 1999) has been deliberately deciding not to intervene in the biomedical sector for the past several decades, in order to avoid a split in the party and a division of the government coalition. Once in power, a "secular coalition" composed of Liberal, Socialist and Green Parties promptly promoted a permissive ART policy. In the agro-food domain, government parties supported GMOs as a promising economic sector up until 1999, when Green parties succeeded in putting the GMO issue on the public and governmental agendas, which ultimately led to a more restrictive regulatory practice. In terms of federalism and administrative responsibilities, GMO policy was decided and implemented in a cooperative way and with a clear sharing of competencies across various levels of power. A strong Biosafety Council ensured, de facto, a multilevel and inter-policies coordination of GMO issues, which clearly favored a stronger state interventionism than in the ART field, where such a vertical cooperation among the federal and regional levels was absent and where a centralizing administration was also lacking. The delegation of decision, implementation and control tasks to various "private agencies" (e.g., National Bioethics Committee, the Physicians' College, Federal Commission on Research on Embryo) also explains why the Belgian ART policy is permissive.

Finally, the pressure for harmonization from *international organizations and European rules* was stronger in the agro-food and environmental sector than in the biomedical sector. Several EU directives on the contained use, deliberate release and commercialization of GMO were transposed into Belgian legislation. This external pressure towards a more restrictive regulation of GMO within the European Union reinforced the impact on the aforementioned national factors. In the ART field, the requirements to sign and ratify the European Convention on Biomedicine are voluntary. Belgium has, as of yet, decided against signing this convention, which prohibits the creation of embryos for scientific purposes. In fact, Belgium adopted a permissive law on this issue, in the hopes of stopping the brain drain and keeping the best researchers at home.

In the following sections, we proceed in two steps to elaborate in further detail our argument: we first present the designing process and content for each biotechnological field; then we explain the two diverging political regulations of ART and GMO in Belgium by comparing them along the three categories of independent variables outlined in the first chapter.

## PERMISSIVE ART POLICY

This biomedical field has a long history in Belgium. Since the 1960s, R. Schoysman has been developing artificial insemination techniques at the Vrije Universiteit Brussels (VUB). In 1983, the first test tube baby in Belgium was born as a result of an IVF performed at the Katolieke Universiteit Leuven (KUL). Today, Intra-Cytoplasmic Sperm Injection (ICSI) is a well-known technique, but what is far less known is that ICSI was first used at the VUB by P. Devroey's and A. C. Van Steirteghem's research team. In 1992, this team reported the first successful birth using this technique. It is, therefore, not surprising that Belgium has one of the highest densities of ART centers in the world, and that the country is facing an "over supply" of providers. As a consequence, there is strong economic competition between the most efficient (i.e., university) ART centers. Today, embryo and stem cell research is also a very promising area of biotechnology, with such potential techniques as replacing tissue damaged by injuries or treating serious diseases such as Parkinson's. Thus, competition between scientists will certainly increase in this area too. In this regard, Belgium has shown its willingness to be an important international player in the biomedical R&D sector by such actions as proposing a permissive declaration on cloning at the United Nations (see A/C.6/58/L.8 of October 17, 2003).

The history of the design process for Belgian ART policy can be divided roughly into four sequences. The first phase takes place from the 1960s until the 1980s, during which time the *development of artificial insemination is self-regulated by physicians*. In 1975 the National Council of the Medical Order introduced Article 88 into its Medical Code of Practice. It stipulated that artificial insemination by donor (AID) was limited to married couples, who must give their written and informed consent, and sperm donors would remain anonymous. The anonymity of the sperm donor remains the official rule. It was never questioned and even reassessed in the media in

2001 by the Minister of Social Affairs, Public Health and Environment. However, the marriage condition was relaxed. The newly formulated principle in the Code of Practice focused on the well-being of the future child, on the required skills of the physicians, and on the role of the ethics committees within each hospital.

The second phase of the policy design covers the period from 1982, when a first attempt to regulate artificial insemination by donor was made at the decentralized level, to the *paternity order protected by a revision of the Civil Code* in 1987. In 1982, a bill was introduced in the Parliament of the French Community, a subnational entity within the federal system of Belgium, for the purposes of regulating sperm donation and conservation, artificial insemination by donor and the question of paternity. The French Community, however, has no formal power to modify the Civil Code, which is de jure necessary for regulating the paternity order. Thus, the bill was never adopted. A *federal law* relating to this bill was eventually enacted in February 1987 in order to modify Article 318 § 4 of the Civil Code. Since then, the husband who gives his written and informed consent to an artificial insemination by donor cannot contest the paternity of his wife's child.

The third phase (1986–1995) is characterized by the *creation of the National Consultative Committee for Bioethics* (NCCB) and the *licensing scheme* for ART centers. The idea for creating such a committee had been discussed at the federal level since May 1984. The law to institutionalize the National Consultative Committee for Bioethics was finally adopted in March 1995. The NCCB, which is mainly composed of physicians and researchers, produced several recommendations for ART regulation, including an official agreement for ART centers (recommendation Nr 6 of the NCCB). This proposition was translated into a decree in February 1999. Since then, specific conditions must be respected by a hospital if it wants to get an official license for practicing ART, as well as an obligation for annual reporting. In fact, this agreement scheme mainly consisted of a sharing of the "ART market" between the competing university hospitals and private clinics. Each of the ART centers party to the agreement were able to practice reproductive medicine according to its own Code of Conduct, as there was still no concrete federal regulation of ART. Apart from this institutional and procedural design, several substantial ART regulations (both restrictive and permissive bills) were also proposed during this phase, but none were accepted due to a lack of consensus among Christian Democrats and secular parties.

The fourth phase began in 1997, denoting the period when the policy agenda focused primarily on the *research on human embryos* and on *cloning*. The developments in this phase are linked to two European initiatives. First, the Council of Europe issued, in April 1997, the *Convention on Human Rights and Biomedicine* (with Article 18 on embryo research), which was then amended with an additional protocol on the prohibition of the cloning of human beings in January 1998. Secondly, the *European Directive 98/44/CEE* (July 1998) based on the legal protection of biotechnological inventions excludes the human body and processes of human cloning from patents and the human embryo from use "for industrial or commercial purposes" (Articles 5 and 6). These two European initiatives contributed to the debate about ART in Belgium. Several bill propositions on embryo research and cloning have been introduced in both the Senate and the Chamber. A Senate special commission on bioethical matters was established in February 2001, which led to the adoption of a permissive federal law regarding research on in vitro embryos and cloning in May 2003.

At first glance, Belgium seems to be a so-called "bioethical paradise" for those who want to practice (doctors) and to have access (patients) to ART with a minimum of restrictions (see table 9.1). Physicians are self-regulated as far as the hospital rules for ART practices are concerned, and the recent law of May 2003 allows them to conduct research on in vitro embryos and therapeutic cloning. The access to ART for patients is also very high. Married, cohabiting couples, single parents and hetero- or homosexual couples

**Table 9.1.  Current ART Policy Design in Belgium**

|  | Research | Practice of fertility treatments | Measures aimed at patients |
|---|---|---|---|
| Policy Design | - Creation of embryos and stem cells for research and therapeutic cloning permitted: authorization delivered by the Commission on embryos research (mainly composed by physicians) <br> - Reproductive cloning prohibited | Licensing of ART centers by regional authorities, with a decentralized self-regulation by physicians (through Local Ethics Committee) <br> - AI, IVF, ICSI and PID permitted <br> - Control of practices quality by the Physicians' College | - Access for patients not restricted <br> - Public funding of IVF and ICSI (if limited number of transferred embryos, in order to avoid the expenses of multiple pregnancies) |
| Ideal Type | Permissive | Permissive | Permissive |

choose the ART center that meets their specific needs. Moreover, the Social Security system has recently broadened the insurance coverage (decree
adopted in April 2003). Medical acts (gynecological practices and medicine)
as well as laboratory work (such as ICSI) are now reimbursed. Considering
both the large autonomy of target groups and the high access of final beneficiaries, the policy design is clearly permissive in comparison to the intermediate ART policies adopted by France, Spain, or the Netherlands, as well
as by very restrictive countries such as Germany, Switzerland or Norway
(Bleiklie et al. 2003).

## INTERMEDIATE GMO POLICY

Belgium also played a pioneering role in the area of research and development for agro-food GMOs. The scientific community attributes the first genetically modified plant to M. Van Montagu (University of Gent), in collaboration with J. Schell (Gent) and H. Goodman of the San Francisco
Medical School. In 1983, these researchers implemented the first vegetal
transgenesis and, subsequently, obtained transgenetic tobaccos resistant to
the antibiotic *kanamycine* (Casse-Delbart 1996: 61). Also, M. Van Montagu
created, along with his collaborators, a spin-off company called "Plant Genetic Systems" (PGS), the success of which has been encouraging to the development of companies in the biotechnology sector in Belgium. Indeed,
PGS is still an acknowledged leader in the field, despite being bought by
Aventis and subsequently Bayer.

In addition to the activities of private companies, several universities continue to invest efforts in GMO, in particular, the Vlaams Interuniversitair Instituut voor Biotechnologie (VIB, Ghent), the Facultés universitaires des sciences agronomiques de Gembloux, the Katholieke Universiteit Leuven
(KUL) and the Universiteit Gent (RUG). In addition, Belgium, and in particular (West) Flanders, possesses a significant concentration, especially considering the size of the country compared to other European states, of companies
in this sector, including Bayer CropScience (formerly Aventis), Monsanto,
AgrEvo (PGS/Hoechst), Advanta SES, Novartis Seeds and Syngenta.

The policy-designing process can be broken down into three stages. During the first stage (1983–1993), the federal minister of agriculture allowed
field trials on a *general legal basis*. He also received recommendations from
the Biosafety Service of the former Hygiene and Epidemiology Institute.
Furthermore, European directives 90/219/CEE and 90/220/CEE took a
long time to be transposed into Belgian law. Firms, ministries and adminis-

trations were able to overcome the legislative gap by anticipating the application of the European decisions. At this time, an important political idea emerged, namely the institutionalization of a Biosafety Council that would consolidate the federal and regional levels of power.

The second stage (1994–1999), characterized by the *institutional consolidation of the so-called Biosafety Council*, produced a specific regulation that was focused on GMOs. The regulation of contained use was assigned to the Regions (decrees of 1993 in Brussels, 1995 in Flanders and 1996 in Wallonia). In 1998, the Biosafety Council and its secretariat (the Biosafety and Biotechnologies Service) were formally appointed and were assigned the role of advising decision-makers for all GMO activities, from contained use to commercialization. From this point on, Belgium possessed a structure for risk assessment and, more generally speaking, scientific expertise.

The third stage (1999–2005) was sharply marked by *public debates and rules reinforcement*. Five factors worked towards further restricting GMO activities. Firstly, in January 1999, the dioxin crisis brought to light the agrosanitary issues, which subsequently led to the creation of a Federal Food Safety Agency (FSA). This new institution was assigned responsibility for, among other topics, GMO controls, thus linking the dioxin crisis to GMO policies. Secondly, a number of NGOs launched important information and mobilization campaigns (e.g., Greenpeace's campaign in 2000 or *Nature et Progrès* in 2002). Related activities, such as the illegal rooting-up of field trials, started to appear in 2000. Thirdly, Belgium participated in debates over the European moratorium, which was imposed in June 1999. The commercialization of GMOs was thus put on hold until May 2004 when labeling and traceability rules were adopted. The moratorium further contributed to the climate of controversy insofar as Belgian political leaders also banned field trials. For the first time in the history of GMO regulation, ministers of public health and environment did not follow the recommendations of the Biosafety Council (April 19, 2002, April 30, 2002, April 11, 2003). Fourthly, responsibility for agriculture was transferred from the federal to the regional level with the creation of the FSA. This institutional change influenced the regulation of GMOs by introducing new actors. Fifthly, these policy actors needed to transpose the EU Directive 2001/18 related to the release and the marketing of GMOs. In February 2005, they succeeded in adopting a firm decision in which the content of this new regulation was quite precise.

In sharp contrast to the biomedical sector, Belgium is certainly *not* a bioethical paradise for private firms that wish to produce and commercialize

GMOs. As a matter of fact, the interventionist Belgian regulation imposes numerous conditions for the contained use and deliberate release of GMOs (see table 9.2). The regional ministers of environment manage the authorization for laboratory research on GMO. At the federal level, the Biosafety Council gives advice to the minister of public health, consumer protection and the Environment, who regulates field experiments, the production and commercialization of GMOs. Finally, the Federal Food Safety Agency is responsible for the *ex post* control of deliberate release and labeling of GMO and GM Food. We may qualify such a policy design as intermediate, in comparison to, on the one hand, the permissive policies implemented by the United States or Canada and, on the other, the restrictive policies adopted by Switzerland or Germany.

## EXPLAINING THE DIFFERENCES BETWEEN ART AND GMO POLICIES

From a comparative point of view, the ART and GMO sectors in Belgium share at least one similarity: decision-makers intervene to regulate biotech-

**Table 9.2.   Current GMO Policy Design in Belgium**

| | Field trials | Commercialization and cultivation | Consumers and environment |
| --- | --- | --- | --- |
| Policy Design | - EU deliberate release directive transposed: authorization required by federal Minister for Public Health and Environment, and regional Minister (portfolio government) consulting the Biosafety Council<br>- Inspections<br>- Penalties | - EU law for GM Food and feed as well as deliberate release<br>- Labeling requirement<br>- Inspection by the Federal Food Safety Agency<br>- Sanctions | - EU law for GM food and feed as well as deliberate release<br>- Labeling requirements defined at EU level<br>- Traceability requirements defined at EU level<br>- Inspection by the Federal Food Safety Agency<br>- Sanctions |
| Ideal Type | Intermediate | Intermediate | Intermediate |

nology sectors that raise economic issues and involve fundamental bioethical questions. Conversely, the empirical analysis of the design processes highlights significant differences in the political treatment of these conflicts of values and beliefs, as well as in the content of the resulting biopolicies, with ART obtaining a procedural and permissive policy and the GMO sector having a substantial and intermediate policy.

In order to explain these diverging biopolicies, we start from the specificities of each biotechnological sector, and first and foremost from the actors who constitute the policy network. After having compared the structures of the networks belonging to the ART and GMO sectors, we look in detail at the arenas and institutional rules that the various policy actors have mobilized, both on the national and international levels, in order to promote their policy position.

## "Policy Community" in the ART-field versus "Issue Network" in the GMO-field

From a methodological point of view, we have identified the main actors of the policy network in the ART sector by analyzing the content of official documents (e.g., bills, laws, parliamentary debates, administrative reports), and conducting interviews with thirteen experts from the political-administrative sector, medical and research sector and voluntary associations. We also applied the so-called "reputation" approach (Laumann and Knoke 1987) to identify the network of actors concerned by the public regulation of ART. For this purpose, we gave the chosen experts a questionnaire consisting of a list of ninety-seven organizations, which had been identified during our preliminary documentary analysis, that were divided into five categories: hospitals, pressure groups, political parties, and public authorities at the federal level and the federated levels. The experts identified twenty-six influential actors among the ninety-seven proposed in the questionnaire. Among these twenty-six actors, the nine most influential ones— according to the weighted scores of their reputation as powerful actors— constituted the core of the policy network. These actors were located in both the political field (the Socialist and Liberal parties, the special Senate Commission on Bioethics) and in the medical field (Cliniques universitaires de Bruxelles - Hôpital Erasme, Akademisch ziekenhuis van de Vrije Universiteit Brussel, Institut de Morphologie-Pathologie de Loverval, Centrum voor Menselijke Erfelijkheid van de Katolieke Universiteit te Leuven, the National Bioethics Committee, of which the key members are physicians and researchers and, finally, the Belgian Register for Assisted Reproduction). Thus, it seems appropriate to focus our explanations of ART policy mainly on the role of political actors (parties, members of parliament) and on medical actors (practitioners, hospitals).

With a similar aim of explaining the resulting GMO policy, we applied the same methodology we had in the ART sector. A reputation analysis enabled us to circumscribe the network of the most influential actors during the decision-making process. After a first documentary analysis, we interviewed eleven experts in the GMO field. These experts were from political-administrative, academic, industrial and associative sectors. We also gave them a questionnaire containing a list of ninety-one organizations divided into six categories: university research centers, pressure groups, private companies, political parties and public authorities at the federal level and at the federated levels. These experts identified thirty-seven influential actors from the ninety-one proposed in the questionnaire. This reputation analysis also indicated that the core of this network was composed of thirteen actors who belonged to political (Green parties: Ecolo and Agalev, cabinet of the minister for public health), administrative (Biosafety Council), industrial (Bayer CropScience, Monsanto, EuropaBio) and academic sectors (Vlaams Interuniversitair Instituut voor Biotechnologie, Facultés universitaires des sciences agronomiques de Gembloux, and Katolieke Universiteit Leuven) as well as to environmental NGOs (Greenpeace, Nature et Progrès) and consumer protection organizations (Test-Achats). This demonstrated that the GMO issue was more broadly distributed across all social arenas, with the notable exception of farmers, whose influence was not highlighted by the reputation analysis. This broad distribution also applied to the thirty-seven other actors forming the global policy network.

The interviews and the reputation analysis allowed us to identify the policy networks of these two sectors, as well as the most influential actors at the heart of each network. On the basis of the typology developed by Marsh and Rhodes (1992: 251), table 9.3 shows that the actors' network of ART looks very much like a "policy community," while the GMO network is of the "issue network" type.

The structures of the policy networks in the ART and GMO sectors are substantially different, located at opposing ends of the continuum suggested by Marsh and Rhodes (1992). This difference in the structure of the policy actors helps us to understand the divergence of the decision-making process in the ART and GMO policy fields. The designing process of ART policy is characterized by a large number of nondecisions, which are the result of a control of the agenda-setting and the decision-making process by a small number of actors, all of them experts in the biomedical sector. This

**Table 9.3.   Types of Policy Networks for the ART and GMO Fields**

| Dimensions of the policy networks | | ART field: "policy community" | GMO field: "issue network" |
|---|---|---|---|
| Membership | Number of participants | *limited*: medical sector (university hospital and clinics), a few political entrepreneurs from socialist and liberal parties and Senate Commission | *large*: private biotech firms and economic associations, environmentalist groups, consumer associations, scientific researchers from universities, political entrepreneurs from green parties and an administrative agency (Biosafety Council) |
| | Types of interests | Professional interests of *physicians and researchers* clearly dominate | Broad range of competing research, economic, environmental and social/democratic interests |
| Integration | Frequency and continuity of interaction | *Frequent and high quality* of interaction between physicians (e.g., through the National Council for Bioethics, the Physician's College) and contacts with the politicians (Senate Commission) especially of their respective sociological "pillar" (counselors) | Access to the network *fluctuates* in time (e.g., sporadic and non-institutionalized actions of environmentalist groups) and contacts fluctuate in frequency and intensity (e.g., crisis periods) |
| Consensus or conflict on values and core beliefs | | *Consensus*: Participants do not share the same values and beliefs (e.g., ontological status of embryo, family model), but they agree to "live and let live" and, thus, they adopt a cooperative "tit for tat" strategy based on a bioethical pluralism | *Conflict*: Participants disagree on basic values and beliefs on the manipulation of living organisms, the usefulness of GMO for agriculture and development aid, the risks of GMO for health and the environment, etc. and they radicalize their policy position |
| Distribution of resources within: | the network | Expertise clearly in the hand of the *physicians* (whom politicians are dependent on) and *little* transparency about ART practices | Competing expertise of biomolecular biologists and other scientific experts (whom politicians are dependent on) as well as divided civil society and transparency about deliberate releases of GMO |
| | the participating organizations | *Hierarchical* through local bioethics committees and organization of (university) hospitals | Loose organization of environmentalists, retailers, and consumers groups |
| Power and decision-making style | | *Balance of power* between the members (sociological pillars), and positive sum game (if the policy community is to persist) | *Unequal* (but fluctuating in time) power of pro and contra GMO, and zero-sum game (defeat of the other side of the issue network as goal) |

*Source*: adapted from Marsh and Rhodes (1992: 251)

"nondecision process" is illustrated by the fact that, since the 1980s, dozens of bills aiming at regulating ART have been proposed in Parliament without being adopted. In contrast, the design process of GMO policy is characterized by various successive and hotly debated decisions with repeated conflicts between social groups during the agenda-setting and decision-making processes.

But, as rightly underlined by Dowding (1995), we cannot immediately conclude that the various network types linearly lead to different policy designs and policy outcomes. To establish such a link, we have to study the concrete interactions between the actors of the network during the decision-making process: "'policy community' and 'issue network' are merely labels attached to an explanation of differences between policy formations in different sectors. The labels do not themselves explain the difference. The explanation lies in the characteristics of the actors" (Dowding 1995:142).

We must, therefore, concentrate on the interests and beliefs of policy actors, as well as on their relationship of cooperative exchange or confrontation. The target groups of the ART policy were clearly physicians and researchers. These actors, even if they did not share the same values and interests, nevertheless decided on a concerted course of action with the aim of limiting all public intervention. The self-regulation of their practices—at the decentralized level of all ART centers that display great bioethical pluralism—was politically acknowledged as a credible alternative to a public debate and regulation that would seek to harmonize ART practices in Belgium. This "corporate" management of the problems raised by ART was never challenged by the actors external to the sector. Two factors favored nondecisions and purely procedural regulations, such as the official recognition of the ART centers since 1999 and the reimbursement of expenses for patients since 2003, that have secured the freedom of biomedical research with the law of May 2003 on in vitro embryo research: first, the lack of mobilization by patients as final beneficiaries and, secondly, the overall positive public evaluation of biomedical applications—for example, genetic tests are supported by more than 90 percent of the Belgians.

The weak presence of administrative actors at the core of the policy network is an additional point to be highlighted. The tasks of conception, implementation and control of the policy are de facto largely delegated to bodies that are mainly composed of physicians, such as the National Bioethics Committee, the Physicians' College, the Federal Commission for the Medical and Scientific Research on Embryos in vitro. Given the quasi-monopolistic position of the medical community in the entire design process, it is hardly surprising that the policy design is "permissive," leaving it to the discretion of each ART center to self-regulate its practices and, consequently, to potentially exploit the legal possibilities of research and therapies according to the law of May 2003.

In contrast, the target groups of the public GMO regulation, most notably the private companies and scientific researchers, find themselves in a markedly different position. The scientific community is more divided today than during recent decades with regard to the expected benefits and potential risks of GMO. Public opinion is also currently very critical of green biotechnologies. In 1999 for example, only a minority of 47 percent of the Belgians supported GM food. Thus, one might also wonder whether the policy design directly mirrored citizens' attitudes.

Private companies and scientific researchers also seemed incapable of proposing a self-regulation for their sector. Moreover, they had to face the strong organization and mobilization of the various end beneficiaries of the policy design, which primarily consisted of environmental and consumer protection associations. With the Biosafety Council and its minister acting as intermediaries, the state thus claimed the role of referee between the two "advocacy coalitions," and of arbitrator between conflicts of interest. The "intermediary" character of the policy design represents a kind of middle way between the positions of the defenders and the opponents of GMO.

This brief analysis of the interaction of actors involved in the ART and GMO sectors stresses the fact that, beyond this structural data on the actors present, the decision-making modes differed greatly. Scharpf (1989) lists three types of potential relationships between the actors of a policy network. They can be "competitive" when the actors have incompatible interests; "indifferent" when they try to avoid dependence on other actors; and "cooperative" when they have common interests and trust each other. These three types of relationships imply three different decision-making styles: "confrontation" between competitors, whose overall goal is the defeat of the other side; "bargaining" between actors, who are motivated by their self-interest but who are unconcerned about the gains of the others; and "problem-solving," when there is a cooperative search for solutions satisfying all the actors. Based on the distinction proposed by Scharpf (1989), the ART sector reflected a problem-solving situation in which a limited number of medical actors adopted a "tit for tat" strategy, despite their diverging bioethical values and beliefs, while the decision-making style in the GMO sector can be characterized by confrontation between various actors, polarized at the heart into two coalitions.

Focusing on the values and interests of the actors, we have briefly demonstrated the links that exist between one type of policy network (policy community versus issue network), one type of decision-making style (cooperative problem-solving versus competitive confrontation) and one type of policy design (permissive ART policy versus intermediate GMO policy). We have not explicitly dealt hitherto with the influence of institutional rules, both at the national and at the international level.

## Party Politics, Multilevel Governance, and Inter-Policy Coordination

With the aim of integrating institutional dimensions characteristic of the Belgian political system, we will now focus the analysis on "party politics," the weight of federalism and, to a lesser extent, the administrative inter-policies coordination.

Political parties played a key role in the regulation of the two biotechnological sectors. The comparison of the designing processes highlighted the decisive weight of the secular coalition (as from 1999) for ART and of the Green parties (as from 1999) for GMO. In the first case, the Christian Democrats, when they were partners in the central coalition, avoided putting bioethical issues on the governmental agenda, mainly for two reasons. The first reason was to protect the internal cohesion of the Christian Democratic Party. The second reason was to preserve the governmental coalitions that have united Catholic and secular parties since World War II. Because the parties generally adopt divergent religious-philosophical guidelines, the coalitions have always been at risk when problems along a religious cleavage appear on the political agenda. For example an unprecedented political and constitutional crisis broke out when abortion was decriminalized in 1990. But, in July 1999, for the first time in fifty-four years, a new secular coalition was constituted by Liberal, Socialist and the Greens. Thus, Socialist and Liberal senators used the arena of the Senate to impose the agenda and the decision on embryo research, which was eventually adopted in 2003. This simple fact is significant in two ways: on the one hand, it was the first time that a political decision filled the legal vacuum regarding biomedicine; on the other hand, the adopted regulation was permissive, as research on embryos and even the creation of embryos for research purposes was allowed.

In the GMO case, it was the ministerial position of representatives of the Green Party (Ministry of Public Health, Consumer Protection and the Environment during the legislature 1999–2003) that was crucial in pushing towards a more restrictive regulation. Before 1999, the Greens were not a member of the governmental coalitions, which had repeatedly supported GMO as a promising economic sector. As soon as they arrived in power, the

Green ministers refused several times to authorize deliberate GMO releases, despite favorable recommendations from the Biosafety Council. Furthermore, the Greens tried to implement the EU Directive 2001/18 in a very restrictive way, even before it was formally transposed into Belgian legislation. However, after the general elections of May 2003, the Green parties returned to the opposition. The new federal coalition (formed by Socialist and Liberals) finally adopted, on February 21, 2005, a decree transposing the EU rules. This Belgian regulation of GMO was less restrictive than the initial proposition of the Greens: the Socialists and Liberals agreed, as a compromise solution, on an intermediate biopolicy. In fact, the Socialists were influenced by the previous GMO policy of the Greens and, to some extent, also defended a quite restrictive GMO regulation. In so doing, they tried to get some support from the Greens' electorate during the regional elections in Flanders. For example, in February 2004, two Socialist ministers of Public Health and Environment refused to authorize the field experiments of a transgenic rape plant in Belgium, while permitting its import from abroad.

In short, the contents of the biopolicies' designs depended on a partisan struggle that, obviously, led to the adoption of a policy design in agreement with partisan ideologies and the electoral interests of their political promoters. On the one hand, the secular parties in government supported a very moderate state intervention with regard to ART, while, on the other hand, the environmentalist parties, once in power, promoted a more interventionist regulation with respect to GMO and, thereby, influenced the subsequent decisions of the Socialists.

Beyond the party-political influence, it has to be noted that the federalist structure of Belgium and its administrative organization also partly explain why the policy designs of the two sectors diverge upon comparison. Federalism has clearly been an obstacle during the decision-making process in the ART sector, whereas it is rather a driving factor in the GMO sector. Belgium federalism combines elements of both jurisdictional and functional federations (Watts 1999; Braun 2000). The distinction between these two types of federations rests upon the method of division of competencies between the federal government and the subfederal governments. In jurisdictional federations, full competencies in specific sectors are attributed to each level of government. In their respective areas of exclusive jurisdiction, governments can act unilaterally in jurisdictional federations. This is the case for GMO, where their contained use depends on the region in question, but the deliberate release of GMO depends on the federal state with the authorization of the regional minister being a prerequisite.

In a functional federation, competencies are attributed along the functions of policy formulation and policy implementation. While formulation is normally the responsibility of the federal government, policy implementation usually belongs to subfederal governments. This is the case for ART: the federal level decides on the programming (e.g., number of agreements for ART centers) and the regions have been delivering official licenses to ART centers since 1999. The federal state also controls *ex post* the quality of care, as a result of the recent creation of the Physicians' College.

In contrast to jurisdictional federations, functional federations normally require intensive intergovernmental cooperation. Yet, in the two biotechnological sectors that we are comparing here, the opposite is true. There is no multilevel coordinating organ for ART, which would be necessary to have a coherent policy, but the Biosafety Council has nearly exclusive responsibility for this task in the GMO sector. The existence of a political-administrative structure surely represents one of the conditions for (at least the implementation of) an interventionist public policy. It is thus not surprising that, in the ART sector, the representatives of the medical sector have appropriated the tasks of implementation and control (e.g., by the Physicians' College) and, hence, contribute to the legitimacy of a "permissive" policy design.

The differences connected to the political-administrative capacity in the ART and GMO sectors tend to reinforce the impacts of federalism on the adopted biopolicies. Indeed, the arena of debate and coordination of the biomedical stakes has been limited to Parliament and especially to the special Senate Commission on Bioethics. This body plays a de facto filtering role for access to the decision-making process, just as it proceeds to sequencing—rather than coupling—the debates on bioethical questions (e.g., decriminalization of euthanasia). This process enhances the capacity of the biomedical practitioners and researchers to influence this body in particular for adopting a policy design that is compatible with their interests. Indeed, physicians are the main experts heard by Parliament.

Conversely, the institutional arenas (e.g., the Biosafety Council, the Federal Agency for Food Safety, Parliament and the municipalities) and extra-institutional arenas (e.g., the media, street demonstrations, the rooting-up of experimental fields) that the opponents of GMO invest in are multiple and varied. This makes the regulation more open and also allows the coupling of GMO with previous agro-alimentary crises (e.g., the mad cow crisis in March 1996; the dioxin crisis in February 1999). Baumgartner and Jones (1993) argue that the existence of many "institutional venues" in a country yields opportunities for "policy entrepreneurs," who use the different venues for strategic policy-making. The potential opportunities for policy entrepreneurs are increasing with the growing number of policy arenas. This hypothesis seems completely plausible in the case of GMO in Belgium: indeed, the multiplication of the arenas and institutional rules mobilized offer the possibility, for the opponents of GMO, to have their point of view heard, and, ultimately, to have an "intermediate" policy design adopted.

While national variables are certainly relevant, the international context also appears decisive to an explanation for the content of the Belgian biopolicies. Thus, we finally address the question of how international regimes have influenced the policy design in the ART and GMO sectors.

## Europeanization by Changing the Structure of Opportunities

The international contexts in which the ART and GMO policy design process have developed are also very different. In the case of ART, there was no institutional pressure to harmonize Belgian policy with supranational norms. The Convention on Human Rights and Biomedicine of the Council of Europe (1997) and the Additional Protocol on the Prohibition of Reproductive Cloning (1998) are to be signed and ratified on a *voluntary* basis by countries. Belgium has not adhered to them. Conversely, in the case of GMO, Belgium *was obliged* and subsequently transposed the European directives on contained use of GM Microorganisms (revised 90/219/EEC: 98/81/EC) and on deliberate release (repealed 90/220/EC: 2001/18/EC). As the other EU member states have also done, Belgium also directly applied the regulations on traceability and labeling (1819/2003 and 1820/2003). This difference seems fundamental in the sense that the domestic actors are *not* capable of withdrawing from this international framework.

The question of embryo and cloning research has been put on the Belgian political agenda partially because of the biomedicine convention. However, its restrictive Article 18 (which prohibits the creation of embryos for scientific purposes) has not been supported by a sufficient majority of votes in favor of signing and ratifying the convention. On the contrary, it has had a triggering effect in the sense that Belgium adopted a national law in 2003 that dealt with this issue, but that ratified a solution that was markedly more liberal than the one initially proposed by the convention. In other words, the Belgian political entrepreneurs made a deliberate choice to first vote for a permissive law (also with the aim of keeping the best researchers in Belgium) and, subsequently, to potentially ratify the convention by expressing their reservations (using Art. 36 of the convention) about Article 18. But we must insist on the fact that Belgium has not yet signed the Biomedicine Convention.

The influence of the European directives is clear in the GMO sector. The process of "Europeanization by changing domestic opportunity structure" (Knill and Lehmkuhl 2002) seems even more plausible since it is happening in a sector where the actors of the policy networks are polarized. According to Héritier and Knill (2001), the potential European impact on national public policies might increase with the extent to which a domestic policy context is characterized by a contested interest constellation and a relatively even distribution of powers and resources across opposing actor coalitions. In such a context, changes in domestic opportunity structures, which are induced by European rules, might be more likely to favor one specific advocacy coalition. Indeed, the EU Directive 2001/18/CE seems to have opened a political "window of opportunity" for the GMO opponents camp, which tried to "instrumentalize" the future 2001/18/CE in order to legitimize their arguments. The final decision (see the decree of February 2005) was adopted while the Greens were again in the opposition and thus does not follow all of their requirements. In any case, the resulting intermediate GMO policy corresponded to a rather restrictive implementation of the EU rules. All in all, in both ART and GMO cases, the supranational norms were translated into national law according to the values and interests of the dominant Belgian policy actors in the sense of either a less (ART) or a more restrictive (GMO) regulation.

Finally, it has to be noted that certain foreign experiments have also been the object of a reappropriation and of a "policy transfer" (Dolowitz and Marsh 2000) by certain Belgian actors. Thus, both the creation of the Federal Food Safety Agency and the organization of a "consensus conference" on GMO at the local level in April 2003 were directly inspired by previous French practices. Indeed, these two processes of "lesson-drawing" have tended to enlarge the debate on GMO by both coupling GMO with food crises and by directly involving citizens in the political debate. These two factors have indirectly contributed, through their strategic valorization and the media attention they have received, to the adoption of a policy design qualified as "intermediate."

## CONCLUSION

As a conclusion of this comparative analysis, we underline the importance of considering all three sets of factors (i.e., the sectors' characteristics, the national institutional arenas, and the international rules) for explaining the permissive policy design in the ART sector versus the intermediate regulation of GMO in the agro-food sector. No one single group of variables is sufficient to explain the different results on its own. Table 9.4 presents an overview of the explanatory factors according to the working hypotheses formulated in chapter 1.

The two Belgian case studies presented in this chapter tend nevertheless to demonstrate that the structuring of the policy actors, at the sector level, lay at the heart of the explanation of the resulting biopolicies. The agenda-setting and decision-making arenas and the institutional rules enabled, even reinforced, the strategies followed by the dominant actors of the policy network. One key example of this was the continued mobilization of the Biosafety Council as a multilevel coordination body in the GMO field versus the creation of private bodies representing the interests of physicians to overcome the absence of an administrative structure in the biomedical sector. A further example is the anticipated application of European GMO directives versus the substitution of the European Convention on Biomedicine by a more permissive Belgian law regulating the research on embryo and therapeutic cloning. In short, a well-founded explanation of the political regulation of biotechnologies in Belgium cannot escape a detailed analysis of the influential actors at the heart of the sector concerned, the interests and values they defend and, finally, the winning strategies they use to obtain a policy design formulated according to their policy positions.

Moreover, it is interesting to note that the GMO controversy led to an intermediate regulation. In contrast, the biomedical regulation, discussed in rather closed circles, led to permissive decisions insofar as it provided for a

**Table 9.4.   Explanatory Factors of Biopolicies in Belgium**

|  | ART | GMO |
|---|---|---|
| Policy Network |  |  |
| H1: Policy community of targets | YES | NO |
| H2: Broad mobilization by critical interest groups in issue networks? | NO | YES |
|  |  |  |
| Country Patterns |  |  |
| H3: Christian Democrats/Green Party in government / key player? | YES | YES |
| H4: Critical groups pushing on national and federated level? | NO | YES |
| H5: Ministries in charge not sharing interests of target groups? | NO | YES |
|  |  |  |
| Internationalization |  |  |
| H6: Europeanization / European policies successfully used by critical interests? | NO | YES |
| H7: Policy transfer through competition? | NO | NO |

large scope of applied techniques and broad access for patients. From a theoretical point of view, this raises questions about the link between the nature of the designing process and the design as policy content. Does an open and conflictual policy formulation process automatically lead to a restrictive rather than a permissive policy content? This is a crucial point in the debate over representative and deliberative democracy and their capacity to deal with risk-related policy issues, especially those related to biotechnologies.

Last but not least, from a normative point of view, ART and GMO regulations highlighted divergent rules of the game. The Belgian ART sector was mainly regulated by procedural and informal rules. The legal constraints were rather scarce or general, as there was no precise law on ART techniques for instance. Self-regulation of practitioners was and remains common practice. Local ethics committees served as safeguards in this respect. Although this model was based on informal rules, it has been strongly respected by the target groups. All ART community members obey self-defined rules that are specified, in a diverging way, at the hospital level. In addition, the patients as final policy beneficiaries also do not dispute it. In stark contrast, GMO activities were framed by very detailed legal prescriptions. European decisions, either as transposed directives or as directly applying regulations, were formally integrated in the legal apparatus. We thus find ourselves faced with a formal but controversial GMO regulation. The institutionalization of the issue has not hampered its conflictive character. As a mater of fact, GMO opponents continue to contest the previous decisions (e.g., through illegal rooting-up of field trials) and try to reinforce the political regulation of green biotechnologies. Thus, all policy actors have not accepted the current policy as legitimate, despite its open and participatory decision-making process. This is another point to be compared among case studies for theoretical generalizations, that is, consensus easier for informal regulation or, in other words, do formalization and institutionalization of policies stoke the controversy around them?

## REFERENCES

Baumgartner, F. R., and B. D. Jones. 1993. *Agendas and Instability in American Politics*. Chicago: University of Chicago Press.

Bleiklie I., M. Goggin, and C. Rothmayr (eds.). 2003. *Comparative Biomedical Policy. Governing Assisted Reproductive Technology*. London: Routledge.

Braun, D. (ed.). 2000. *Public Policy and Federalism*. Aldershot: Ashgate.

Casse-Delbart, F. 1996. La transgénèse végétale. In *Les plantes transgéniques en agriculture. Dix ans d'expériences de la Commission du Génie Biomoléculaire.* Ed. A. Kahn. Pp. 55–88. Montrouge: John Libbey Eurotext.

Dolowitz, D., and D. Marsh. 2000. Learning from Abroad: The Role of Policy Transfer in Contemporary Policy-Making. *Governance: An International Journal of Policy and Administration* 13 (1): 5–24.

Dowding, K. 1995. Model or Metaphor? A Critical Review of the Policy Network Approach. *Political Studies* 43 (2): 136–58.

Héritier, A., and C. Knill. 2001. Differential Responses to European Policies: A Comparison. In *Differential Europe: New Opportunities and Restrictions for Member-State Policies.* Eds. A. Héritier et al. Pp. 257–94. Lanham: Rowman & Littlefield.

Knill, C., and D. Lehmkuhl. 2002. The National Impact of European Union Regulatory Policy: Three Europeanisation Mechanisms. *European Journal of Political Research* 41 (2): 55–80.

Laumann, E., and D. Knoke. 1987. *The Organisational State: Social Choice in National Policy Domains.* Madison: University of Wisconsin Press.

Marsh, D., and R. A. W. Rhodes. 1992. Policy Communities and Issues Network. Beyond Typology. In *Policy Networks in British Government.* Eds. D. Marsh and R. A. W. Rhodes. Pp. 249–69. Oxford: Clarendon Press.

Scharpf, F. W. 1989. Decision Rules, Decisions Styles and Policy Choice. *Journal of Theoretical Politics* 1 (2): 149–76.

Watts, R. L. 1999. *Comparing Federal Systems.* Kingston: Institute of Intergovernmental Relations.

## 10

# ARTS AND GMOS IN SWEDEN: EXPLAINING DIFFERENCES IN POLICY DESIGN

*Francesca Scala (Concordia University)*

**B**iotechnology in Sweden has become a site of political struggle as well as a symbol of scientific and technological progress. In the late 1970s, Sweden was among the first countries to launch a public debate on the challenges and merits of biotechnology. Public opinion on biotechnology, however, differs regarding GMOs and ARTs. While the Swedish public has a very optimistic view of embryo and stem cell research, it is highly critical of genetic engineering in the area of agriculture. Government's response to these different biotechnology fields also differs. The policy design of ARTs in the area of research is among the most liberal in Europe while government policy in the area of GMOs has been quite restrictive. This chapter examines the development of policy in the areas of assisted reproductive technologies and genetically engineered organisms. It explores the factors that contributed to diverging policy responses in the two areas paying close attention to factors related to internationalization, the polity and policy networks.

## THE POLICY DESIGN OF ARTS IN SWEDEN: ASSISTED INSEMINATION AND IN VITRO FERTILIZATION

Infertility treatments in Sweden, specifically assisted insemination (AI) and in vitro fertilization (IVF) are governed by two pieces of legislation: the

Insemination Act of 1985 and the In Vitro Fertilization Act introduced in 1989 respectively. Before 1985, donor insemination (D. I.) was conducted without any legal restrictions worldwide (Lalos et al. 2003). Secrecy accompanied this practice in an effort to safeguard the anonymity of both the donor and the couple undergoing the treatment. With the 1985 act, Sweden became the first country to exercise direct statutory control over the practice of D. I. The introduction of the act was motivated by two factors: the well-being of the child conceived through donor insemination; and resolving the legal status of the social father. The act effectively resolved any legal ambiguity by assigning parental rights and responsibilities to the social father. Moreover, the Act requires that D. I. be practiced only in hospitals under the supervision of a physician specializing in reproductive medicine. It also limits access to women who are married or in a long-term, cohabiting relationship.

One of the defining features of the act, which continues to set it apart from legislation in other countries, is the elimination of donor anonymity. Concerned with the "welfare of the child," the Swedish government decided that children born of sperm donation had the right to know his or her biological fathers at an "age of maturity," which has typically been interpreted as age eighteen (Jonnson 1997). During the preparation of the new legislation, the issue of donor anonymity was raised as recently released studies on adoption and expert advice from social workers revealed that adopted children benefited from knowing their biological background. These findings were directly relevant to children born from donor insemination who were refused access to their biological heritage because of the principle of anonymity. Many physicians practicing D. I. at the time opposed the legislation, arguing that it would make it difficult to find potential sperm donors (Sverne 1990). By 1997, the number of children born of donor insemination significantly dropped to approximately fifty to sixty per year from the prelaw rate of 250 per year. The drop could also be attributed to other factors, such as the greater use of alternative treatments and the pursuit of treatments in neighboring countries by couples who favor donor anonymity (Jonnson 1997).

The infertility treatment of in vitro fertilization was first introduced in Sweden in 1978, with the first child born in 1983. In 1989, the government introduced the In Vitro Fertilization Act, which outlined a number of restrictions and prohibitions. As with the case with donor insemination, IVF was to be available only to married women or women in stable, cohabiting relationships. The treatment could only make use of the husband's sperm and the wife's ovum for the act prohibited the use of donated ova and sperm in IVF. The rationale behind these prohibitions was that using donor ova and sperm contradicted the human biological process and "it is so much of an artifice that it could harm the notion of humanity" (Jonnson 1997: 89). In 1995, opinions regarding donated ova in IVF were beginning to change as reflected by the Report of Sweden's National Council of Ethics that recommended the donation of ova should be allowed and governed by the same provisions governing donor insemination (e.g., access to information of donor). Today, there is a move towards allowing donor ova in IVF treatments, however, it has yet to be ratified.

## Delineating the Boundaries of Access

The Insemination Act of 1985 and the In Vitro Fertilization Act of 1989 outlined a number of restrictions on access to infertility treatments based on their definition of infertility and bias in favor of heterosexual couples. Defining infertility in medical terms, both acts stipulate that access to treatments should be restricted to couples whose infertility stems from an underlying medical problem rather than social conditions. Therefore, single women or lesbians experiencing "social infertility" because of their sexual orientation or marital status could not access the treatments because they did not conform to the traditional, heterosexual model of the family (Ross and Landstrom 1999). Other restrictions, such as the upper age limit for women accessing the technologies, is not governed by statutory law but is determined by guidelines in different county councils. In practice, the upper age limit is generally set between thirty-five to thirty-seven years.

In recent years, the issue of access to infertility treatments has been revisited. A number of events have prompted the Swedish government to reevaluate whether lesbian couples should have access to treatments like D. I. In 2001, a legal quagmire emerged when a lesbian separated from her partner sought and won financial support for her child from the sperm donor. Since hospitals are prohibited from treating women who are in lesbian relationships, lesbians are often compelled to pursue sperm donation through private arrangements. In a private arrangement, the sperm donor cannot relinquish financial and legal responsibilities for the child as stipulated in the 1985 Insemination Act. The man in this case was not afforded this legal protection and was therefore regarded as the child's legal father. This case, along with recent changes to the adoption laws, has prompted the Swedish government to change the law governing access to donor insemination. In June 2002, the Swedish government overwhelmingly voted to allow gay couples to adopt children, affording them the same rights as heterosexual couples.[1] Under the new law, gay and lesbian couples registered in a legal partnership are able to adopt children both within the country and from abroad. Along with gay adoptions, the Swedish government also proposed to allow lesbians to access donor insemination within the hospitals. In March 2005, Swedish legislators announced that they would pass an amendment to the law in July 2005 to allow lesbians who live together as registered partners to access assisted conception treatments at state-run fertility clinics.

## Government Control and Limiting Professional Autonomy

Before the introduction of the 1985 Insemination Act, fertility specialists had a great deal of autonomy in delineating the appropriate standards of practice. While accountable to their hospitals and relevant professional bodies, physicians practicing assisted insemination and in vitro fertilization were relatively free to determine issues of access and practice. In 1985, this autonomy was significantly circumscribed as these decisions were brought under the purview of the Swedish government through statutory law. The act placed prohibitions or restrictions on access to infertility treatments. These restrictions were not surprising given that these treatments were being funded by Sweden's publicly funded health care system.

The two acts introduced in the 1980s essentially assigned regulatory responsibility to Sweden's National Board of Health and Welfare. It is the central administrative agency that oversees general supervisory responsibilities, such as publishing guidelines and ensuring that clinics comply with relevant legislation. All fertility clinics are required to submit annual reports on their treatment activities and results to the board. The board's Centre for Epidemiology then prepares an annual report that is made available to the medical profession and the general public. Information on specific clinics, however, is not made available because the public is perceived as having "great difficulties in interpreting these data and that the public could easily be misled" (Nygren 2002: 351).

The In Vitro Fertilization Act of 1987 effectively put an end to professional self-regulation in the area of infertility treatments. Recent legislation passed in January 2003 also places restrictions on certain practices. For example, in the past, fertility clinics enjoyed some level of autonomy in determining the maximum allowable limit of embryos that are transferred during an IVF treatment. This was left to the discretion of individual clinics and practioners; however, common practice was two or three embryos. Concerned with multiple births, the recently passed legislation calls for the transfer of only one embryo or two for exceptional cases during an IVF cycle. Administered by Sweden's National Medical Board, the regulation applies to both public and private IVF clinics. Other facets of reproductive technologies are also governed by statutory legislation. For example, Swedish legislation allows for the cryopreservation of embryos, oocyte and sperm for up to five years. Certain forms of micromanipulation, such as ICSI and assisted hatching are allowed while cytoplasmic transfer is prohibited (Jones and Cohen 2001).

Overall, Sweden's policy design on reproductive technologies, specifically those practices related to infertility treatments, can be characterized as intermediate with a few notable exceptions (see table 10.1). In terms of access, infertility treatments such as assisted insemination and in vitro fertilization are available only to couples deemed to be in a stable relationship. Single women and lesbian and gay couples are currently unable to access these services, although this will change with the introduction of new legislation in July 2005. Prenatal genetic testing and certain forms of micromanipulation, such as assisted hatching, are allowed under certain conditions. The policy design of reproductive technologies also contains some restrictive features. For example, current legislation prohibits certain practices, such as embryo and oocyte donation and the use of donor sperm in IVF treatments.

A central feature of Sweden's policy design is its reliance on statutory law and the concomitant limitations it places on the professional autonomy of the medical community in determining standard practices of care in the area of infertility treatment. As mentioned earlier, legislation rather than professional guidelines regulate the practice of assisted insemination and in vitro fertilization. This has led to tensions between the National Board of Health and Welfare and medical practitioners and fertility clinics. For example, the Swedish government's decision to abolish the principle of donor anonymity was greatly contested by medical professionals who were concerned that it would scare off potential donors. The government's primary concern was the rights and well-being of children conceived by artificial insemination rather than the professional concerns of infertility specialists. The professional autonomy of the medical community is further constrained by limitations placed on the number of embryos to be transferred during IVF treatments. Here, the health risks and financial costs of multiple births, a common occurrence with IVF treatments, prompted the Swedish government to regulate, via legislation, the choices available to infertility specialists in treatment protocols.

## THE POLICY DESIGN OF EMBRYO RESEARCH
## AND RELATED PRACTICES

While Sweden's policy design on infertility treatments can be characterized for the most part as intermediate, the country's policy design on embryo and stem cell research is considered to be the most permissive in Europe. While there is no legislation that specifically targets stem cell research, other legislation applies such as the In Vitro Fertilization Act of 1988 and the Act Concerning Measures for Research or Treatment Involving Fertilized Human Ova of 1991. The 1988 act allows experimentation on embryos within fourteen days of fertilization for the purpose of improving infertility treatments. It also places some restrictions and prohibitions on certain practices, such as the genetic modification of embryos and the implantation of a research embryo in a woman. The 1991 act regulates reproductive research and research on embryonic development, covers storage of embryos and allows for embryos to be cryopreserved for five years (European Parliament 2000).

In the early 2000s, the Swedish Medical Research Council held internal discussions on the matter of stem cell research and somatic cell nuclear transfer (i.e., therapeutic cloning) and began devising guidelines. The following year, policy preparations were transferred to a newly created funding organization, the Swedish Research Council (*Vetenskapsrådet*). In December 2001, the council published guidelines that reaffirmed Sweden's established policy of permitting research on human embryos. While the council recommended prohibiting reproductive cloning and the creation of embryos from eggs and sperm for research purposes, it did, however, permit the creation of embryos by somatic cell nuclear transfer, deeming it to be "ethically justifiable." In January 2002, the Swedish government upheld the council's views and took steps to change its laws in order to make these practices legally permissible. In February 2005, the Swedish government voted in favor of research on human embryos for the purpose of isolating embryonic stem cells and performing therapeutic cloning.

The decision to allow therapeutic cloning with only minor restrictions reflects the Swedish government's long-standing commitment to reaping the economic rewards of biotechnology for the country's economy. Indeed, an important objective of the newly created Swedish Research Council is to promote biotechnology research and as well as "Swedish participation in international research cooperation" (Swedish Research Council 2001: 2).

Moreover, Sweden's policy on embryo and stem cell research was largely driven by the scientific community, with little involvement of the public in the decision-making process. While the issue received coverage by the major Swedish newspapers and other media, there was no formalized system of public hearings and consultations (Mounce 2003).

Today, biotechnology, especially stem cell research, has become a leading industry in Sweden. The country has the fourth largest biotechnology industry in Europe with 235 biotechnology companies—the largest number of biotechs per capita worldwide. The country hosts several of the world's leading research institutes in this area, including the publicly funded Karolinski Institute, whose researchers have developed twenty-four of the forty-four stem cell lines that were discovered worldwide. One of the defining features of biotechnology in Sweden is industry-academic collaboration. Consistent with the Swedish model of "middle way," consensual politics, biotechnology is characterized by close links between industry, universities and the publicly funded health care system. In 1996, 44.5 percent of biotechnology firms were involved in cooperative arrangements with universities and/or government research institutes. Cross-border collaboration among the three sectors is more common in Sweden than anywhere else in the world, manifesting itself in numerous technology transfer units in universities and research institutes, like the Karolinski Institute, which provides a "knowledge bridge" between industry and the universities. Favorable intellectual property rights and a growing venture capital market have also aided in the growth of Sweden's biotechnology sector.

Cross-border collaboration in the area of biotechnology is a function of several factors. First, the universities and public research institutes in Sweden and elsewhere were not shielded from the austere economic policies introduced by governments in the 1990s. Faced with diminished government resources, public universities and research institutes looked to industry for alternative sources of funding when conducting their research work. Second, the commercial and economic potential of biotechnology also appeals to the Swedish government, which is striving to foster long-term and sustainable economic growth. Permissive policies on human biotechnology and greater public-private partnerships are deemed necessary for Sweden to maintain its leading position in what is quickly becoming an increasingly competitive and international biotechnology industry.

The following table summarizes the policy design of reproductive treatments and research in human biotechnology.

**Table 10.1.   Current ART Policy in Sweden**

|  | Research | Practice of Fertility Treatment | Measures Aimed at Patients |
|---|---|---|---|
| Policy Design | - Experimentation allowed on 14-day-old embryos<br>- Therapeutic cloning allowed<br>- Implantation of research embryos in women not allowed | - Publicly funded; private clinics available<br>- Embryo; oocyte donation; and donor sperm prohibited in IVF treatments (under review)<br>- One embryo transferred during IVF<br>- Sperm donor anonymity abolished<br>- Prenatal genetic testing allowed under certain conditions | - AI and IVF available to stable couples (under review)<br>- Lesbian couples recently allowed to access ARTs |
| Type | Permissive | Intermediate | Intermediate |

## REGULATING GMOS IN SWEDEN

While biotechnology in the area of medicine is widely supported in Sweden, its application in the agro-food sector has met with strong resistance from the general public and various interest groups, including consumer-rights and environmental groups as well as some industry organizations. While Sweden has one of the largest and fastest growing biotechnology industries in Europe, it is the least supportive of GMOs among European countries (Durant et al. 1998). This opposition is reflected in the country's restrictive policies on the development, growth and marketing of GMOs. This section examines the policy design of Swedish regulation on GMOs and explores some of the factors that contributed to restrictive policies.

### The Policy Design of GMOs in Sweden

In Sweden, genetic engineering in the area of agriculture is regulated by EC directives. These directives have been incorporated into Swedish law, specifically provisions of the Environmental Code and the Genetically Modified Organisms Ordinance (Minister of the Environment 2005). These provisions govern three areas: (1) contained use, which refers to or-

ganisms that do not come into contact with the general public; (2) deliberate release, involving organisms that are not contained but are planted outside; and (3) placing the organism on the market. The purpose of these
rules is "to protect human health and the environment and to ensure that
special attention is paid to ethical concerns in connection with activities involving GMOs" (Sweden 2005). The provisions require that an investigation
must be undertaken in any of the three areas of activities involving genetically modified organisms.

Different government agencies are responsible for overseeing various elements of agricultural biotechnology. They are: the Swedish Board of Agriculture (SBA), which oversees the deliberate release of GMOs into the environment; the Swedish Environmental Protection Agency, which evaluates
and monitors the environmental impact of GMOs; the Swedish National
Food Administration, which assesses and controls GM food; and the Gene
Technology Advisory Board, whose mission includes monitoring technological developments and informing the public on related issues (Achen 2004).
Assessments of notification concerning GMO release is done in consultation among the Swedish Board of Agriculture, the Environmental Protection Agency and the Gene Technology Advisory Board.

The central administrative body overseeing the deliberate release of
GMOs in the environment, such as field trials, is the Swedish Board of Agriculture (SBA). Since 1989, approximately 92 consents were given for the
deliberate release of GMOs into the environment for reasons other than
placing on the market, such as research and development purposes. Two
applications were not approved (Sandstrom and Norgren 2003). The SBA
did approve one application for the placing of GM potatoes not intended
for food use on the market (European Union 2002). A large majority of the
approved applications were for field trials for potatoes and oilseed rape.
The SBA is responsible for inspecting field trials every year to ensure isolation and proper destruction of crops.

Regulations governing the supervision, penalties, charges and appeals re-
lated to the placing on the market of GM foods are found in Sweden's Food
Act, in particular sections 24–28, 29a, 30 and 35 (SFS 1971: 511). The Food
Act is a frame law that oversees food composition, handling, labeling and
placing on the market. The principle authority in the area is the National
Food Administration, whose identifiable goals are: "safe foods of high quality; fair practices in the food trade; and healthy dietary habits" (SAS 2004).
It is responsible for the premarketing assessment of GM food, which is then
circulated for comments to other government agencies, universities, and organizations representing consumers, retailers and the food industry. The

board is also responsible for inspecting approximately thirty to forty large food producing establishments under the Food Decree (SFS 1971: 807). Until recently, there were no applications requesting authorization for a GM food in Sweden. However, in January 2004, the National Food Administration approved the selling of a beer made with GM ingredients in the Swedish market.

## Ethical Concerns and the Precautionary Principle

In 1994, the Swedish Parliament passed the Gene Technology Act (FSFS 1994: 900), which was based on the EU directives on deliberate release and contained use. The objective of the legislation was the protection of human health and the environment and the incorporation of "ethical concerns" when deciding on the use of GMOs. According to Karlsson (2003), this feature of the Swedish legislation sets it apart from other legislation in Europe. "Although the EU legislation was based on the idea of harmonizing national legislation, the demand to consider ethical aspects was specific for Sweden and had no corresponding provision in the directives" (Karlsson 2003: 53). Through a licensing system, the Gene Act required that potential risks of GMOs be preassessed before a permit could be issued for deliberate release or placing on the market of GMOs. Permits would only be issued if they are deemed "ethically justifiable." The competent authority to oversee the licensing procedure for GMOs would be the Swedish Board of Agriculture. Another ordinance established the Gene Technology Advisory Board to monitor developments in gene technology and provide advice to the government. It was responsible for conducting assessments of GMOs on "ethical or humanitarian grounds."

In 1999, the Swedish government replaced a number of acts on environmental protection with a framework policy, the Environmental Code. EC directives on GMOs were transposed into the legislation. The Gene Technology Act was eliminated and issues related to GMOs were subsequently regulated by the Environmental Code. Most of the provisions are found in chapter 13 of the code. The incorporation of EC directives into the existing Environmental Code Act has significant implications for the system of risk management that governs GMOs. The act introduced the precautionary principle as the central norm for environmental regulation in Sweden. The act states:

> Anyone who carries out, or intends to carry out an activity or action of whatever nature, shall undertake protective measures, follow restrictions and un-

dertake such precautions in general to prevent or counteract that the activity or action in question results in damage or inconvenience with respect to human health or environment. (SFS 1998: 811)

This interpretation of the precautionary principle contained in the Environmental Code Act is the most risk averse in Europe. It "holds that uncertain risk requires forbidding the potentially risky activity until the proponent of the activity demonstrates that it poses no (or acceptable) risk" (Wiener and Rogers 2002: 321). It requires proponents of GMOs to prove their product or experiment will not harm the environment or human health, essentially shifting the burden of assessing risks from the government to the operator of an activity: "the authorities do no have to demonstrate that a certain impact will occur; instead, the mere risk (if not too remote) is to be deemed enough to warrant protective measures or a ban on the activity" (Westerlund quoted in Lofstedt et al. 2002). Regarding GMOs, the definition of risk has been broadly defined to include ethical, social, environmental and health-related concerns.

However, risk assessment and the application of the precautionary principle varies across competent authorities governing GMOs. As mentioned earlier, different government agencies oversee different categories of organisms and their uses. They are individually responsible for regulating the requirements that are to be met by the activities that fall under their supervisory authority and examining the information that is submitted to them by users. The dispersion of authority among various agencies has led to different interpretations of the precautionary principle in the risk assessment processes governing the contained use and deliberate release of GMOs as well as placing onto the market GM products. The Swedish Board of Agriculture (SBA), for example, has been much more lenient in its application of the precautionary principle and its evaluation of risk. Karlsson's (2003) examination of the implementation of the environmental code in the area of deliberate release found that during the period between 1999–2000, the Swedish Board of Agriculture refused to issue a permit to just one of twenty-one applications, which included GM potato with altered composition of starch and containing antibiotic-resistance genes, and herbicide-tolerant sugar beet. These approvals were based on established scientific assessments stated by the board's own regulations.

The SBA's assessment procedure differs from that of the Gene Technology Advisory Board, which centers on ethical concerns. These different assessment procedures and principles have at times come into conflict with one another. For example, in one case involving GM oilseed rape, the Gene

Technology Advisory Board recommended that the application for deliberate release be rejected on ethical grounds. The SBA, on the other hand, recommended that the application be approved based on its own science-based assessment process. As Karlsson explains: "neither obvious stakeholders, such as companies and farmers, nor the more vaguely specified interested parties, such as future generations and other species were identified. The board had omitted to analyze the social benefits and costs for deliberate release" (2003: 57).

While the Swedish Board of Agriculture seems to use a more liberal interpretation of the precautionary principle than that adhered to by the Gene Technology Assessment Board, in recent years it has significantly decreased the number of approvals for field trials. For example, the Board gave permits to only five different GMO crops for the 2001 planting season. The area of land used for GMO has dropped from 353 hectares in 1998 to 22 hectares in 2001 (SBA 2001).

## The Politicization of GMOs

The issue of GMOs became a matter of public debate in Sweden when a genetically manipulated soya product from the United States was set to enter the European market in the fall of 1996. The Swedish Board of Agriculture had reviewed the application and had agreed to permit the American company to import its product before the formal voting procedure of the EU member states (Bauer and Gaskell 2002). However, given that regulatory responsibilities in the area of GMOs are shared by various government agencies and departments, discussions had to be held with other relevant ministers. The Ministry of the Environment opposed the proposal and after three days of deliberation, Sweden would reject the application. According to Bauer et al. (1998) the decision to reject the import of the soya product was not driven by scientific evidence, the criteria dictated by EU directives; rather it was due to the increasing politicization of the issue of GMOs. This event prompted the government to revise the statute dealing with applications for approving the release of GMOs in order to allow it to express its views before the formal voting stage. The Swedish government would later accept the EU's approval of the import of the generically modified soya product (Bauer and Gaskell 2002).

The case of the GM soya product and a similar case involving Ciba-Geigy genetically modified maize inspired a debate on GMOs in the Riksdag and in the general public. Several political parties in Sweden, most notably the Centre Party (the former Agriculture Party) and the Green Party, expressed their opposition to GMOs, citing environmental and health reasons. They called on the government to ban the import of GM foods by invoking Article 16 of the EC Directive on Deliberate Release on the grounds that their health effects and impact on the environment were unknown. At the very least, the group called for the creation of a regulatory regime that would oversee the labeling of all GM foods, and thus protecting the consumer's right to make an informed decision. As Bauer and Gaskell explain, "Parliamentary opponents insisted that they identified with, and were representing, the consumer's position" (299).

Government also solicited the views of several NGOs, including environmental and consumers groups as well as several of Sweden's food distributors and retailers. Environmental organizations like Greenpeace and the Society for Nature Conservation were concerned with the impact of GMOs on biodiversity while consumer groups raised concerns regarding the health risks of the increased use of pesticides and the possible spread of antibiotic resistance. Retailers and consumer groups were in favor of a labeling system that would help individual consumers make informed decisions. In general, opponents, including MPs from the Centre Party and the Green Party, helped frame the debate on GMOs in terms of environmental and health risks, thereby expanding the constellation of actors in the policy network. In the end, however, the critical voices did not prevent the government from allowing the import of the GM soya in Sweden. The Swedish government, led by the Social Democrats, advanced an optimistic discourse on GMOs in general. The government espoused the benefits of genetic engineering and argued that the potential harms of GMOs could be contained within a regulatory framework. The minister of agriculture refused to invoke Article 16, arguing that there was insufficient evidence showing the potential risks of the product and that the issue was moot given that it was impossible to grow soya in Sweden (Bauer and Gaskell 2002). Regulatory restrictions rather than prohibitions were regarded as more appropriate tools for governing genetic engineering.

Today, Swedish policy governing the deliberate release of GMOs and placing on the market of GM goods can best be characterized as intermediate and restrictive, respectively. Just like some other countries in Europe, Sweden has approved a limited number of GM field trials on a case-by-case basis over the years. Regarding the placing of GM products on the market, Sweden has in place a very restrictive policy. Indeed, it is one of several countries that pushed for more stringent traceability and labeling requirements at the European Union level.

Table 10.2 summarizes the policy design of GMOs in Sweden.

Table 10.2.   GMO Policy Design in Sweden

|  | Field trials | Commercialization and cultivation | Consumer and environmental protection |
|---|---|---|---|
| Policy Design | - Field experiments allowed on a case-by-case basis | - GM foods not allowed until recently <br> - GM Beer approved for market | - Traceability and labeling requirements defined by EU Directives |
| Type | Intermediate | Restrictive | Restrictive |

## ART AND GMO POLICY IN SWEDEN: EXPLAINING POLICY DESIGN DIFFERENCES

Human and nonhuman biotechnology has garnered different public and governmental responses in Sweden. In the area of human biotechnology, we can discern an intermediate and permissive policy design governing assisted fertility treatments and embryo research respectively. Agriculture biotechnology, on the other hand, is a mix of intermediate and restrictive policies, depending on the activity (i.e., deliberate release or placing on the market). What accounts for this variation? The following discussion outlines some of the factors that contributed to different policy designs in these areas.

### Policy Network Dynamics

In the case of Sweden, the most important explanatory variables for variations across and within the two fields of biotechnology relate to the type of policy networks that are in place. The policy network governing assisted conception techniques comprises a small number of actors, namely the National Board of Health and Welfare and the medical profession. The latter is represented by two principle organizations: the Swedish Society of Medicine (SSM), which is primarily concerned with issues related to research and education in the area of medicine; and the Swedish Medical Association (SMA), the medical profession's trade union body. In Sweden, the state has traditionally enjoyed a considerable level of autonomy vis-à-vis the medical profession when devising health policy. While still engaging in consensual politics, the Swedish government has traditionally taken the lead in reforming aspects of the health care system (Immergut 1992). As Garpenby explains, "During the 20th century, the relationship between the state and the medical profession has changed, above all because of the all-inclusive, publicly financed health care systems which was successively introduced in Sweden, where terms have increasingly come to be controlled by political decision-makers" (Garpenby 1999: 415). While the state regularly interacts with the medical profession in the area of health policy, it can initiate and change policy without the involvement of interest groups. The autonomy enjoyed by the National Board on Health and Welfare is encouraged by the inability of the medical community to act collectively and speak with one voice on health care matters due to internal divisions and the competitive relationship between the SSM and SMA (Garpenby 1999).

Moreover, due to economic downturns and challenges facing the public sector in general, the Swedish government has been concerned with improving the delivery of services within the context of budgetary constraints. In the area of health care, this has manifested itself in a more active interest in the outcome of clinical medical activity, the central sphere of activity of the medical profession. Within this context, legislation limiting the number of embryos transferred during IVF treatment can best be understood as an attempt to curtail health care costs associated with multiple pregnancies.

The policy network that governs embryo and stem cell research is characterized by strong, collaborative relationships between the scientific community, the pharmaceutical industry and government officials. Sweden has a strong biotechnology industry, especially in the area of pharmaceuticals and is therefore highly susceptible to the competitive pressures of the international market. From the very beginning, industry interests were taken into account by the research community and the Swedish government when fashioning policies in this area. Groups critical of embryo research on religious or moral grounds do not take part in the policy network. The general public as well does not have viable avenues to influence policy on this issue. Instead, the Swedish government has relied heavily on the expert advice of scientists and research councils, as reflected in its adoption of the Swedish Research Council's pro-stem cell research stance. Consistent with the hypothesis suggested at the beginning of this book, this policy community, which adheres to the grand narrative of the new knowledge economy, resulted in a relatively permissive policy design in the area of embryo and stem cell research.

The policy network that governs GMOs is larger and more open than that which exists for human biotechnology. As mentioned earlier, the regulatory regime is divided along different supervisory authorities, basing their decisions on different processes of risk assessments. In the case of the deliberate release of GMOs, the Board of Agriculture rules the day, approving many applications for field trials based on their science-based assessment, which at times overrides the Gene Technology Advisory Board's recommendations that are based on broader ethical concerns. The number of actors in the dominant GMO policy network is more than that of the human biotech sector. It comes closer to an issue network. Government agencies regularly seek comments from key organizations, including environmental organizations, farmers, consumer groups, and organizations representing the food industry. Support for GMOs among these groups is very low. For example, in a 2000 survey only 12 percent of farmers in Sweden would consider growing GMOs (ATL: 4 Feb 2000). The food industry and retailers have also adopted a critical stance by "attempting to prevent the introduction of GMOs into grocery stores and agricultural fields and animal feed" (Akesson 2002). Finally, environmental organizations and consumer groups have been active participants in the debate on GMOs and have been successful in bringing media and public attention to the issue. The Swedish Consumer Coalition, for example, lobbied the government and business to adopt a comprehensive GM food labeling system. Critical of biotechnology, this issue network was successful at obtaining from government restrictions in Sweden's GMO policy. In contrast, industry actors in the area of agriculture biotechnology are fewer in number and have been less vocal in espousing the merits of GMOs in public debates. However, the Swedish Board of Agriculture has acted in favor of agri-biotech firms, only refusing to approve two applications for field trials between 1998–2002.

The type of relationship between the relevant competent agencies and stakeholders also contributed to a restrictive policy design in agriculture biotechnology. For example, the decision to transpose the EU directives into Sweden's existing Environmental Code and Food Act had implications for the type of principles and requirements that would guide risk assessments of GMOs. As mentioned earlier, the Environmental Code requires that the use of GMOs not harm human health and the environment and that it be "ethically justifiable." Consequently, environmental concerns rather than industry interests figure more prominently. Consumer interests are the primary concern of the National Food Administration, which is mandated to ensure the safety of the food supply. The representation of consumer interests and environmental concerns in these agencies led to a more open and contested policy process, and more restrictive policies in agriculture biotechnology.

## Country Patterns

Party politics and public opinion may offer other explanatory variables for differences in policy designs in the two areas. Public debate on embryo research and reproductive treatments played itself out in the daily newspapers. By most accounts, however, the debate in Sweden did not resemble the acrimonious debates that took place in other countries, such as the United States. This is largely due to the fact that the issue of abortion has not been a divisive one in Sweden since the majority of Swedes have a liberal attitude on the matter. As Evers explains, in Sweden, "the right to use contraception and the right to have an abortion are not political issues and are considered fundamental rights [and] the rights of the unborn are not posited in the law on abortion" (Evers 2002: 1580). Moreover, the media, in particular the influential *Dagens Nyheter* has become an important useful outlet for both scientists and policymakers to exalt the benefits of human biotechnology.

The Swedish Parliament also did not play a significant role in framing the issue of biotechnology in Sweden. Political parties have not galvanized public debate on the issue nor have they provided institutional avenues for successful resistance. As expected of a progressive party in the area of ART, the Social Democratic Party, which has been in power during the different periods of legislative action on biotechnology, has favored a more permissive policy design. In the past years, the Social Democratic government has strived to create a favorable political and economic environment in order to attract foreign investment and encourage economic growth. It regards biotechnology as a key industry that has the potential of providing high economic growth and employment levels.

Other political parties, including the Christian Democrat Party, the Green Party, and the Central Parties, are critical of both human and non-human applications of biotechnology, and favor more restrictive policies. During parliamentary debates, the Green and Centre parties called for a complete ban on embryo research. However, according to Kulawik (2003), the parliamentary debate did little to alter the packaging of biotechnology as an economic imperative. She points to discursive and rhetorical strategies on the part of techno-optimists to delegitimize critics by characterizing them as "irrational" and "conservative."

> Although supporters of new procedures emphasized the dangers of "misuse," for example human cloning, they presented their own arguments as progressive and accused the "others" of being backward and enemies of research. (Kulawik 2003: 19)

In recent years, opposition on the part of political parties to issues such as stem cell research has waned. For example, the Swedish government's approval of stem cell research was not vehemently opposed by the Christian Democrats, a party that has in the past voiced its opposition to embryo research. The party stated that it was favorable to stem cell research in principle, although it advocated some restrictions be imposed, such as prohibiting the creation of embryos for research purposes.

Opposition to biotechnology in Sweden is much more pronounced in the area of agriculture. The general public in Sweden is among the most critical of agricultural biotechnology in Europe. A survey in 1999 revealed that 76 percent of Swedes oppose or slightly oppose genetically modified foods (Bauer and Gaskell 2002). Regarding political parties, the Green Party and the Left Party are most critical of agro-biotech. As an environmentalist party, it is not surprising that the Green Party has adopted the most extreme anti-biotech position, calling for an outright ban on the selling of GM foods and the production of GM plants. The Left Party is also critical, asking for a five-year moratorium on agriculture biotechnology. Stakeholders representing consumer organizations and environmental groups forged important alliances with the Green and Left parties in their efforts to bring about stringent policies on GMOs.

However, as the ruling party in Sweden, the Social Democratic Party's more moderate position on biotechnology in general has helped keep in place a case-by-case approach to the risk assessment of GMOs. Several factors may have contributed to the party's more favorable stance. First, like in the area of medicine, the government saw the economic potential of the application of biotechnology in the area of agriculture. Biotechnology was and continues to be regarded as a symbol of economic and technological progress and governments strive to make their mark in an increasingly competitive and global market. Prohibitions would have hampered Sweden's international standing in this new global market. Second, prohibitions on GMOs may have impelled a discussion on the merits and safety of biotechnology in the area of medical research. This would have threatened the economic prospects of the pharmaceutical and other biotechnology industries in Sweden. Third, in the case of the genetically modified soy product from the US, the Swedish government may have considered the ramifications of refusing to allow the import on Sweden–US trade relations and newly developing collaborations in the area of biotechnology. As Bauer explains, "Sweden is highly dependent on exports from its large and successful industrial sector and therefore reluctant to harm relations with the USA" (Bauer and Gaskell 2002: 302).

## Internationalization

Global economic pressures significantly influenced the permissive policy design of embryo research in Sweden. As mentioned earlier, Sweden is a leading country in the area of biotechnology, with strong pharmaceutical and biomedical R&D industries. Biotech firms are numerous and growing, helping Sweden to achieve a "significant presence within global biotechnology networks" (Lofgren and Benner 2003: 29). The Swedish government's decision to allow embryo research and therapeutic cloning was greatly influenced by its desire to maintain and strengthen the international competitiveness of the country's biotechnology industry and foster foreign investment in the domestic economy. In Europe, Sweden hosts the fourth largest biotech industry, assuming a leading position in biotechnology publications and number of biotechnology firms (Hayrinen-Alestalo and Kallerud 2004). It is not surprising that Sweden is feeling the pressure to maintain its competitive edge in the global biotechnology industry. Similar global economic pressures are not present in the area of assisted conception techniques and related practices.

While international economic competitiveness was an overriding concern of Swedish policy-makers in the area of embryo research, certain restrictions found in the legislation comply with guidelines set out by the European Union and other international bodies. For example, prohibitions on the creation of embryos for research purposes are consistent with Article 18.2 of the 1998 European Convention of Human Rights and Biomedicine, which prohibits the creation of embryos solely for research purposes. Moreover, restricting research on fourteen-day-old embryos reflects guidelines established by the international scientific community. Conversely, Swedish policy on reproductive technologies, in particular its restrictive criteria for accessibility, contradicts European and United Nations conventions that affirm the right of all individuals to found a family, including lesbian and single women. These restrictions, however, are expected to be abolished.

In the case of GMOs, multilevel governance is a significant explanatory variable for Swedish policy. As mentioned earlier, EC directives have been transposed into existing legislative and regulatory frameworks. However, it is important to note that Sweden was one of the first countries to enter the public debate on GMOs in Europe and it had already established national legislation preventing the planting of GMOs before it entered the EU. Moreover, it has been instrumental in pressuring the EU for more stringent regulations in agriculture biotechnology, especially in the area of labeling.

The EC directives and moratoriums have also had a deleterious effect on the agro-biotech industry in Sweden. The agro-biotech industry is significantly smaller and less lucrative than the human biotech industry. The sector comprises only eight companies, totaling approximately 600 million SEK in revenues. As mentioned earlier, the number of field trials being conducted in Sweden has steadily decreased. This steady decline, according to some observers, can be attributed to reluctance on the part of companies to pursue GMO development. According to Achen (2004), "One obvious reason for this is the unclear situation regarding the future of GMO in Europe caused by the de facto moratorium on the approval of new genetically modified crops and foodstuff in Europe"(Achen 2004: 148).

## CONCLUSION

In Sweden, a number of factors contributed to differences in policy design between agriculture and human biotechnology. In the former, multilevel governance and a policy network characterized by a fragmented regulatory regime and broad associational system allowed critics of GMOs to influence policy in the area. However, as mentioned earlier, Sweden had already put in place stringent policies before its entry into the EU. In the area of ARTs, assisted conception techniques and biotech research are characterized by different policy designs. In the case of embryo research and related practices, industry interests, the international economic competitiveness of the Swedish biotechnology sector and the influence of the scientific community were all factors that contributed to a more permissive policy design. The Social Democratic Party's commitment to the pharmaceutical-biotechnology sector as well as the long-standing collaborative relationships among the state, industry and the university-research community have contributed to permissive policies in this area. Conversely, the policy network that governs assisted conception, characterized by a weak associational system and strong government agencies, has led to legislative controls on standard practices of care. These findings are largely consistent with the hypotheses presented in chapter 1.

## NOTE

1. It was passed by the Parliament by 198 votes to 38, with 71 abstentions.

# REFERENCES

Achen, T. 2004. Actors, Issues and Tendencies in Swedish Biotechnology. In *Mediating Public Concern in Biotechnology: A Map of Sites, Actors and Issues in Denmark, Finland, Norway and Sweden*. Eds. M. Hayrinen-Alestalo and E. Kallerud. Pp. 113–55. Oslo: Norwegian Institute for Studies in Research and Higher Education.

Akesson, L. 2002. Sweden. In *Biotechnology: Educating the European Public—Final Report*. European Commission, contract HPRP–1999–00007.

ATL [Swedish Farm Journal]. February 4, 2000. www.atl.nu/

Benner, M., and U. Sandstrom. 2000. Inertia and Change in Scandinavian Public-Sector Research Systems: The Case of Biotechnology. *Science and Public Policy* 27 (6): 443–54.

Bauer, M., and G. Gaskell, 2002. *Biotechnology: The Making of a Global Controversy*. Cambridge: Cambridge University Press.

Durant, J., et al. 1998. *Biotechnology in the Public Sphere: A European Sourcebook*. London: Science Museum.

European Parliament, Directorate General for Research. 2000. *The Ethical Implications of Research Involving Human Embryos, Final Study*.

Evers, K. 2002. European Perspectives on Therapeutic Cloning. *New England Journal of Medicine* 346: 1576-79.

Garpenby, P. 1999. Resource Dependency, Organised Medicine and the State: Quality Control in Sweden. *Social Science & Medicine* 49: 405–24.

Government of Sweden. 2002. Invest in Sweden Agency (ISA), Biotechnology.

———. 2004. The Swedish Research System. www.sweden.se/templates/FactSheet____3925.asp (accessed on August 12, 2004).

Hayrinen-Alestalo, M., and E. Kallerud. 2004. Introduction: Towards a Biotech Society—Nordic Perspectives. In *Mediating Public Concern in Biotechnology: A Map of Sites, Actors and Issues in Denmark, Finland, Norway and Sweden*. Eds. M. Hayrinen-Alestalo and E. Kallerud. Oslo: Norwegian Institute for Studies in Research and Higher Education.

Immergut, E. 1992. The Rules of the Game: the Logic of Health Policy-Making in France, Switzerland, and Sweden. In *Structuring Politics: Historical Institutionalism in Comparative Analysis*. Eds. S. Steinmo, K. Thelen, and F. Longstreth. Pp. 57–89. New York: Cambridge University Press.

Jones, H. W., and J. Cohen. 2001. Worldwide Legislation. In *Textbook of Assisted Reproductive Techniques: Laboratory and Clinical Perspectives*. Ed. D. K. Gardner. London: Martin Dunitz.

Jonnson, L. 1997. Regulation of Reproductive Technologies in Sweden. In *Governing Medically Assisted Human Reproduction*. Ed. L. Weir. Toronto: Centre of Criminology, University of Toronto.

Karlsson, M. 2003. Ethics of Sustainable Development: A Study of Swedish Regulations for Genetically Modified Organisms. *Journal of Agricultural and Environmental Ethics* 16: 51–62.

Kulawik, T. 2003. Gender Representation and the Politics of Biotechnology in Sweden: Explaining Liberal Regulations in a Social Democratic State. Paper presented at the European Consortium for Political Research, Marburg, Germany.

Lalos, A., K. Daniels, C. Gottlieb, and O. Lalos. 2003. Recruitment and Motivation of Semen Providers in Sweden. *Human Reproduction* 18 (1): 212–16.

Lofgren, H., and M. Benner. 2003. Biotechnology and Governance in Australia and Sweden: Path Dependency or Institutional Convergence? *Australian Journal of Political Science* 38 (1): 25–44.

Lofstedt, R. E., et al. 2002. Precautionary Principle: General Principle: General Definitions and Specific Applications to Genetically Modified Organisms. *Journal of Policy Analysis and Management* 21 (3): 381–407.

Minister of the Environment. The Swedish Environmental Code. www.miljo.regeringen.se/english/pdf//env-code_resume.pdf.

Mounce, M.-A. 2003. Stem Cell Research in Sweden. BIOTECH Magazine.

Nygren, K. 2002. The Swedish Experience of Assisted Reproductive Technologies Surveillance. World Health Organization *International Reproductive Health*, 32.

Ross, B., and C. Landstrom. 1999. Lesbian Identity and Organizing. In *Women's Organizing and Public Policy in Canada and Sweden*. Eds. L. Briskin and M. Eliasson. Montreal: McGill-Queens University Press.

Sandstrom, A., and L. Nogren. 2003. Swedish Biotechnology: Scientific Publications, Patenting and Industrial Development. *Vinnova*, the Swedish agency for innovation systems.

Sverne, T. 1990. Biotechnology Developments and the Law in Changing Family Patterns. *International Social Sciences Journal* 126: 465–93.

Swedish Board of Agriculture. At www.sjv.se//genteknifoltforsok/faltforsok.htm (accessed August 12, 2005).

Swedish Research Council. 2001. *Guidelines for Research-Ethical Review of Human Stem-Cells Research*.

The Swedish Gene Technology Advisory Board. www.sjv.se/net/GMO/GMO+-+English/Gene+technology+Advisory+Board (accessed August 15, 2004).

Weiner, J. B., and M. D. Rogers. 2002. Comparing Precaution in the U.S. and Europe. *Journal of Risk Research* 5 (4): 317–49.

# ⑪

# SWITZERLAND: DIRECT DEMOCRACY AND NON-EU MEMBERSHIP—DIFFERENT INSTITUTIONS, SIMILAR POLICIES[1]

*Christine Rothmayr (Université de Montréal)*

Switzerland is a rather unique case from an institutional point of view within our set of countries. It is the only European country under study that is not a member of the European Union, and its political system combines direct democracy with federalism and a form of government that is neither parliamentary nor presidential. Given that the direct democratic instruments and the supranational context are different from our other countries, one might wonder whether Swiss policies also differ considerably from those of neighboring countries. For example, one might speculate that Swiss policies would be more or less restrictive than in EU member states. The analysis reveals, however, that while the designing processes are structured by the peculiarities of Swiss institutions, the resulting policies have considerable commonalities with those of neighboring countries in terms of restrictiveness. Policies for genetically modified organisms in the agro-food sector (GMOs) are similar to those of EU countries with a restrictive implementation and transposition of EU norms, such as Germany or Great Britain. The practice of assisted reproductive technologies (ART) is similarly restrictive to neighboring countries, like Italy, Germany or Austria, but clearly more restrictive than in a great number of other countries, like France, Belgium, Spain, or the UK. Yet, the restrictive policies in both fields are not the result of policy transfer through competition or learning from neighboring countries. With the exception of embryonic stem cell research, international competition plays a limited role. As the following analysis argues, it is rather the broad mobilization of critical interest groups, negative public opinion with respect to certain applications, the framing of biotechnology as potentially harmful technology, the unusual coalition between political parties from the left and the right of the spectrum, and the lead of administrative units not necessarily in favor of the interests of the target groups that together explain the restrictive policies for ART and GMOs in the Swiss case.

## COMMON ROOTS OF REGULATING RED AND GREEN BIOTECHNOLOGY

The Swiss people have so far been called to vote five times on legislative projects related to biotechnology, and will vote a sixth time in the near future. First of all, in 1992, they adopted a constitutional article on biotechnology and ART with 73.8 percent in favor, as a result of a popular initiative entitled "Against the abuse of biotechnology and assisted reproductive technology" (Beobachterinitiative). The constitutional article established the basis for adopting federal laws and ordinances in both fields, hence legislation in the ART and GMO fields have common roots.

With respect to ART, a federal law and two ordinances were designed on the grounds of this constitutional article during the 1990s (FmedG: SR 814.90, VNEK: SR 814.903, FMedV: SR 814.902.2). The designing of the law was influenced by a second popular initiative, "Initiative for procreation respecting human dignity" (Initiative für menschenwürdige Fortpflanzung) sponsored by opponents of the 1992 constitutional amendment. It aimed at reversing the constitutional amendment of 1992 by fully prohibiting IVF and insemination by donor. This time, the Federal Council did not make a direct counter-proposal, but proposed a federal law on the grounds of the existing constitutional article as an indirect counter-proposal. Parliament adopted in December 1998, after a very conflictual debate, the Federal Law on ART and recommended the rejection of the initiative, which the people effectively did in 2000, with 71.9 percent voting against. The third popular vote on ART-related issues, was, however, just around the corner. As a result of technological progress and because the current policies only incompletely addressed the question of embryo research, by the end of May 2002, the federal government sent a proposal for a law on embryo research into the preparliamentary consultation procedure. During parliamentary debate, this proposal was transformed into a Federal Law on Stem Cell Research and the regulation of embryo research in general was postponed to the future Federal Law on Human Subject Research. Against the "Federal Law on Stem Cell Research" a facultative referendum[2] was launched by left and religious-conservative groups, independently from each other. A facultative referendum allows for, after the collection of fifty thousand signatures, a challenge to federal laws adopted by Parliament, which must then be submitted to a popular vote. The law was accepted in 2004 (StFG SR 810.31, VStFG SR 810.311).

Direct democracy played an equally important part with respect to green biotechnology. After the adoption of the constitutional article, as in the case for ART, a second popular initiative, the "Gene-Protection-Initiative" (Volksinitiative 'zum Schutz von Leben und Umwelt vor Genmanipulation, Gen-Schutz-Initiative) influenced decision-making. The initiative aimed at prohibiting genetically modified animals, the deliberate release of GMOs, patenting of genetically modified animals and plants or parts thereof, and procedures to create them. In addition, the initiative asked for strictly regulating the use of gene technology in the nonhuman field. The initiative was rejected with 66.7 percent voting against it in 1998. In parallel to the initiative, existing norms were adapted to address green biotechnology. The revision of the "Federal Law on Environmental Protection" starting in 1995 introduced authorization requirements for GMOs (deliberative releases, putting on the market) from 1997 onwards.[3] Declaration duties were introduced for GM food, and an authorization for placing products on the market has been required since 1995.[4] Because these legal revisions left regulatory gaps, but also in answer to the "Gene-Protection-Initiative," in 1997 Parliament adopted the "Gen-Lex" motion asking the Federal Council to prepare a legislative project that would comprehensively address the issue of GMOs in the nonhuman field. The Federal Council sent a project into preparliamentary consultation in 1997 and instituted the Federal Ethics Commission for Gene Technology in the Nonhuman Field (i.e., before the vote on the initiative took place in 1998). The rejection of the initiative in 1998 opened up the road to adopt the Gen-Lex project. The Federal Council presented its project to Parliament in 2000, proposing to revise and update existing legislation. Parliament, however, preferred one main federal law addressing gene technology in the nonhuman field instead of the revision of existing legislation. The Federal Law on Gene Technology (GTG SR 814.91) was finally adopted in 2003.

The struggle between critics and promoters of gene technology, however, continued. A third popular initiative was initiated in 2003, the "Initiative for Foods from Gene-Technology-Free Agriculture" (Eidgenössische Volksinitiative 'für Lebensmittel aus gentechnikfreier Landwirtschaft'), proposing a five-year moratorium on the use of genetically modified plants, seeds and animals in Swiss agriculture. The initiative passed in November 2005 with 55.7 percent support, becoming only the fifteenth popular initiative to be approved since 1891.

With respect to supranational norms, in the field of ART, in 1999, Switzerland has signed, but not yet ratified, the Convention on Human Rights and Biomedicine plus additional protocols. In contrast, Switzerland signed and ratified the Cartagena Biosafety Protocol in 2002. In order to comply with the protocol, an ordinance was adopted, namely addressing the question of the export of GMOs not regulated by the other laws and ordinances (Cartagena-Verordnung, CartV, SR 814.912.12)

## RESTRICTIVE POLICIES FOR ART AND GMO

In both fields, ART and GMOs in the agro-food sector, the Swiss design corresponds to the restrictive ideal-type of design. Over time, however, the design of public policies in both fields has moved into a moderately less restrictive direction with, in the case of ART, the adoption of the Federal Stem Cell Law, and changes in the implementation practice, in the case of GMOs in the agro-food sector.

The Swiss ART design aims at securing the well-being of the child in general, at protecting human dignity, the personality and the family and also wants to prevent the abuse of ART. These goals correspond to the restrictive Swiss policy design. Federal policies strongly limit the autonomy of the medical community by prohibiting a number of techniques such as egg and embryo donation; preimplantation diagnostics; cryopreservation of embryos;[5] surrogate motherhood; genetic engineering on gametes; germ cells and embryos; therapeutic and reproductive cloning; as well as chimera and hybrid building. To produce an embryo solely for research purposes is prohibited and stem cells can only be derived under specific conditions from leftover embryos. The question of embryo research in general, however, has so far not been comprehensively regulated. For the techniques that are not fully prohibited, the policies define in detail under what conditions and how doctors are allowed to practice them and under which conditions research projects on embryonic stem cells might be approved. In fact, the design prescribes licensing requirements, defines medical indications and how certain techniques have to be practiced, proscribes inspections and controls, specifies reporting duties and formulates information and counseling requirements towards patients. Access to ART is limited to heterosexual couples, and, for certain techniques, further limited to married couples; access also depends on the financial means of the patients to cover the respective costs of treatment (see table 11.1).

With respect to stem cell research, this brief description reveals that Switzerland does not belong to the group of the ideal type of restrictive countries that fully prohibit stem cell research (cf. Austria, Ireland) or limit it to imported stem cell lines (Germany). Furthermore, sperm donation is not prohibited. Access is only limited to married couples in the case of IVF with sperm donation, and otherwise ART remains accessible to nonmarried stable couples. Nevertheless, the Swiss design prohibits egg and embryo donation as well as preimplantation diagnosis and, in some cases, limits access to married couples and therefore still comes closer overall to the *restrictive type* of design.

The Swiss design for GMOs in the agro-food sector also corresponds to the *restrictive* ideal-type of design due to the restrictive authorization practice. The policy goals in the field of green biotechnology are similar to the ones for ART. The policies aim at protecting humans, animals and the environment from the misuse of biotechnology and at securing a use beneficial to humans, animals and the environment. The Swiss design with respect to GMO is strongly based on the precautionary principle. Contained use, deliberate releases, production and distribution of GMOs, and products containing GMOs, are submitted to strict procedures of authorization, labeling (with a 1 percent threshold)[6] and traceability, guaranteeing free

**Table 11.1.   Current Swiss ART Policy Design**

|  | Research | Practice of fertility treatments | Measures aimed at patients |
|---|---|---|---|
| Policy Design | Creation of embryos for research purposes prohibited; research question not yet fully addressed; derivation of stem cell lines under strict conditions from leftover embryos permitted | PID, embryo, and egg donation prohibited; transfer of embryos limited to three | Access limited to stable couples, for some techniques to married couples |
| Type | Intermediate | Restrictive | Restrictive |

choices and transparent information to consumers. In addition, until November 27, 2010, there is a moratorium on genetically modified plants, seeds and animals in Swiss agriculture. Imports, however, are not affected by the moratorium (e.g., the import of genetically modified soybeans by the food industry). Furthermore, the new Swiss liability requirements in the GMO field are, according to experts, much more far reaching than provisions in other countries. To genetically modify vertebrates for commercialization in the food market is also prohibited.[7]

The implementation of federal law has so far been particularly restrictive with respect to deliberate releases. Since the introduction of authorization requirements with the revision of the Environmental Protection Law in 1995, no deliberate release of genetically modified crops had been authorized until 2003. After a long court battle however, a small experimental release for research purposes by the Swiss Federal Technical Institute was authorized in 2004. Thus, already before the adoption of the five-year moratorium, the authorization practice for deliberate releases corresponded to a de facto moratorium. Hence, in Switzerland no commercial deliberate releases have taken place so far or will take place until 2010. There is also hardly any commercialization of novel foods because the two main food distributors decided in 1999 to ban all products containing GMOs from their shelves. However, four authorizations for GMO crops (maize, soy) have been granted and several demands are pending; it remains to be seen how the authorization practice for deliberate releases develops in the near future.

The implementation is organized differently in the two sectors of ART and GMOs in the agro-food sector. In the case of the ART practice, and in accordance with the Swiss type of "implementation-federalism," the main *implementers* are the cantons, whose leeway is very limited, as the federal law and the ordinances predefine the details of implementation. The implementation of the Law on Stem Cell Research, however, is a federal matter. The Federal Office of Public Health is in charge of authorization procedures and the Swiss National Advisory Commission on Biomedical Ethic has to approve the submitted research projects.[8] In contrast, the implementation of green biotechnology policy pertains to the federal level. Several federal administrations are involved in implementing GMO policies (see table 11.2) in the agro-food sector. The Federal Office of Public Health is in charge of implementing norms in the field of GM food and additives. It consults with the Committee on Biological Security, the Environmental Protection Agency, the Federal Office for Agriculture, and the Federal Veterinary Office before authorizing a product. With respect to deliberate re-

**Table 11.2. Current Swiss GMO Policy Design**

|  | Field trials | Commercialization and cultivation | Consumer and environment |
|---|---|---|---|
| Policy Design | Only one single scientific trial authorized; de facto moratorium on field trials | De facto no commercialization of GM food, despite authorization of some foods; no cultivation of GM plants, no transgenetic animals (moratorium); very strict liability measures | Traceabiltiy and labeling of products mandatory (threshold 1%), GMO-free Swiss agriculture (moratorium), imports permitted |
| Type | Restrictive | Restrictive | Restrictive |

leases, the Environmental Protection Agency takes the lead and consults with the same administrative units and committees already listed for GM food including the Office for Public Health. In addition, the relevant canton is consulted, and the Federal Ethics Commission in the nonhuman field issues its position in the matter under scrutiny. For contained use a special unit has been created to coordinate the requests, the Contact Bureau for Biotechnology. Depending on the type of request, different federal offices are consulted, in addition to the ones already mentioned, namely the National Accident Agency with respect to workers' security.

## BROAD MOBILIZATION, CRITICAL PUBLIC OPINION, AND SELF-REGULATION WITHOUT IMPACT

In both policy fields, we find a broad mobilization of interest groups, namely around the various initiatives and referendums. The decision-making process and the final design were in both cases strongly influenced by interests critical towards biotechnology. Promoters of green and red biotechnology, however, successfully mobilized against propositions for total prohibitions of basic applications and research as proposed by the "Initiative for procreation respecting human dignity' and the 'Gene-Protection-Initiative," yet were defeated in the case of the moratorium for GMOs in Swiss agriculture. Hence, interest groups critical of GMO and ART as well as the target groups (i.e., industry, medical, and research interests) realized some of their goals. The interest groups critical of GMO obtained a strict framework and the target groups managed to reject extensive total prohibitions for medical, pharmaceutical and industrial research and applications.

According to the results of the reputational approach and the documentary analysis, the ART policy-designing process was overall an open one, giving access to interests voicing strongly diverging policy preferences. Within the critics of ART, who were promoting very restrictive policies and total prohibitions of basic techniques and research on embryonic stem cells, we find religious interest groups, on the one hand, and environmental interest groups, on the other hand, displaying an overall critical approach not only to reproductive technologies in particular but gene technology in general. Organizations representing the disabled were also opposed to ART. Patient organizations participated only marginally in the policy-making process. Feminist interest groups were generally critical towards ART, but were divided about whether to fully prohibit all techniques or not (Rothmayr 2003). In contrast to the abortion issue, their influence on the ART issue remained limited (Moroni 1994). With respect to the main target group, medical practitioners and researchers were not opposed to state regulation in general, but favored a moderate state intervention. Representatives of the medical interests had successfully challenged total prohibitions on the cantonal level by calling upon the Swiss Federal Supreme Court. Although this decision and the lobbying of medical and research interests on the federal level contributed to preventing total prohibitions, they did not succeed in realizing their policy preferences for a moderate intervention, namely to allow for egg donation, preimplantation diagnostics, to guarantee the anonymity of the donor and to allow for therapeutic cloning. They were not successful in realizing their preferences because of the broad mobilization from both the left and the religious-conservative spectrum. The mobilization went beyond the usual health and research policy networks and the medical and research interests did not enjoy privileged access to participating in the policy-making process mainly for two reasons. First, ART was not primarily seen and discussed as a health policy issue. The "Beobachter" initiative framed the issue in terms of "protection against abuse" and thereby reflected the committee's concerns with technological progress and its impact on society. This initial frame continued throughout the designing process. Secondly, the Department of Justice and not the Office of Public Health had the lead in this matter (see below). The mobilization and the framing also contribute to explaining why the self-regulation of the medical community had no preempting effect on the state intervention and why the current policies are considerably more restrictive than the former self-regulatory policies. Beginning in 1981 the Swiss Academy of Medical Sciences

(Schweizerische Akademie der Medizinischen Wissenschaften, SAMW) had issued standards for self-regulating the practice of ART in Switzerland (SAMW 1981, 1985 and 1990). They were effectively respected in practice, and many clinics even limited their offers more than required (e.g., by only offering ART with the gametes of the couple). On the cantonal level we can observe a pattern that corresponded to translating self-regulation into governmental regulation. On the national level, however, we do not observe any impact on the problem pressure or the framing of the issue. To the contrary, the early, substantial and effectively implemented self-regulation resulted in more restrictive governmental policies because of the mobilization of critical interest groups and the lack of privileged access for the medical interests.

With respect to GMOs in the agro-food sector we have found a few parallels to the ART policy network. While the degree of mobilization varied over time, the two camps (i.e., the biotechnology promoters and the biotechnology critics) remained basically the same throughout the different stages of decision-making. The resistance to gene technology emanates from the left, green and feminist sections of the political spectrum. The anti-gene technology coalition consists of environmental groups, animal rights groups, consumer groups and some farmer organizations (i.e., well-established interest groups promoting their traditional goals) (Graf 2003: 229). The different interest groups coordinated their activities from the 1990s on through an umbrella organization, the SAG, the Swiss Working Group on Biotechnology (Schweizer Arbeitsgruppe Gentechnologie). By launching the "Gene-Protection-Initiative" critics generated agenda-setting effects, leading to the Gen-Lex Motion and the Federal Law on Gene Technology, and created considerable pressure for restrictive solutions and strict implementation practices. Hence, the two principal supermarket chains in Switzerland, Migros and Coop, banned all GM food and products from their shelves in 1999. They argued that consumers did not want this type of product. By practically eliminating the market for GM food, they indirectly strengthened the anti-GMO interest group's coalition. Economic and industry interest groups, in particular the pharmaceutical industry and the agro-food industry,[9] as well as the research community, mobilized against the initiative and in support of an internationally compatible regulation of Gene Technology in Switzerland. In connection with the Gene-Protection-Initiative, the mobilization was unusually strong among researchers in the public and private sectors. They feared a marginalization of Switzerland as a place of research if the initiative would be accepted. Since the rejection of the Gene-Protection-Initiative we can generally observe less mobilization among interest groups and the issue has become less salient in the media. For the elaboration of the Federal Law on Gene Technology the coalitions remained relatively the same; positions became, however, more nuanced because the questions to decide upon were now much more detailed. It is interesting to note that agricultural interests are now even more clearly allied to consumer interest groups than before, because of the various agrofood scandals of the last five years. In the case of the latest popular initiative that instituted a five-year moratorium, the farmer associations supported the initiative.

This brief description of the actor network revealed that both camps successfully realized some of their goals. The interest groups critical of technology attained their goals with respect to deliberative releases, labeling and traceability requirements, the strong institution of the precautionary principle in the current legislation and the five-year moratorium. The pro-gene technology groups achieved their goals with respect to red biotechnology issues, namely an efficient implementation of authorization procedures for contained use and the admission of certain GM food products.

In both fields, actor strategies were influenced by public opinion that, along with media coverage, was linked to and influenced by the main events of the decision-making processes (i.e., the initiatives and the respective voting campaigns) (see, for example, Bonfadelli and Dahinden 2000; Eisner, Graf, and Moser 2003). Opinion polls revealed that the evaluation of biotechnology varied considerably according to the field of application. At the risk of oversimplifying, we can say that medical applications (medication, genetic testing) were positively evaluated, whereas genetically modified crops and plants, GM food and the cloning of animals was generally rejected. In comparison to the EU average, the Swiss more strongly rejected GMOs in the agro-food sector, and their rejection was particularly strong when it came to GM food (Bonfadelli 2000: 94; Eurobarometer, GfS 2003). Public attitudes have certainly influenced the policy design, directly through the rejection of the initiative in 1998, but also indirectly. In the case of the Gene-Protection-Initiative the strategies of the pro-gene technology interest groups successfully framed the issue in terms of the utility and potential progress of red biotechnology and possibly also contributed to strengthening the positive attitudes towards red applications. With respect to the recent GM Food Moratorium Initiative, the fact that GM food was strongly rejected by the Swiss population entered political strategy by limiting the target of this new initiative to GM food and products and—in contrast to the earlier Gene-Protection-Initiative—not touching upon medical, pharmaceutical and industrial applications.

## PARTY POLITICS AND DIRECT DEMOCRACY

The structure of the Swiss political system helps to explain why both sides were partially successful in realizing some of their goals and why the search for compromises dominated the preparliamentary and parliamentary stages in both fields. The use of the popular initiative instrument had three types of effects: first, it put pressure on the political agenda to address ART and, in the case of GMOs, to close legislative gaps; second, it was used as a "pledge" in negotiating policy solutions in the preparliamentary and parliamentary stages (Verhandlungspfand; Linder 1994; Papadopoulos 2001) and third, it was used to adopt a moratorium, thus passing restrictive policies that had failed to gain a majority in Parliament.

In the case of ART, the Beobachterinitiative was a motor of innovation leading to addressing the issue on the federal level by adopting the constitutional article. The rather restrictive content of the initiative was fully integrated into the counter-proposal of the Federal Council in the parliamentary debate. The political parties agreed with the Federal Council that the issue needed to be regulated on the federal level. They were, however, in favor of integrating the restrictive and substantial clauses proposed by the initiative into the counter-proposal, while the Federal Council wanted to limit the constitutional article to stating that ART was a federal matter. Only a very small minority in Parliament, mainly the Liberal Party, argued that the state should not intervene at all. The Radical Party, and also a number of members from the Social Democrats, the Christian Democrats and the People's Party—all four represented in government and having together a large majority in Parliament—wanted certain restrictions, but without a total prohibition of basic techniques. At the same time, opinions within the Social Democrats, the Christian Democrats and the People's Party were divided, and total prohibitions were also supported or considered by members of these three parties. Furthermore, some small Center and Left parties, such as the Protestant Party, the Greens and Feminists, supported a total prohibition of IVF and donor insemination. In short, the vast majority wanted to set clear limits to ART. Based on different beliefs, Christian religious beliefs, on the one hand, and a skeptical attitude towards technical progress and science in society in general, there was also considerable support for total prohibition from both the Right and the Left sides of the political spectrum (Amtliches Bulletin SR: 1990: 477–93: 1991: 250–457, 615; NR: 1991: 556–67, 588–636, 1288, 1408). In order to build a viable majority, a design needed to be found, that satisfied the concerns of those members of Parliament, in particular of the governmental parties who were considering total prohibition, should the constitutional article turn out to be too permissive. Such a majority would not only serve to pass a constitutional article that did not include total prohibitions of IVF and gamete donation, but it would also allow for broad support by the political elite in order to enhance the chances of the counter-proposal being accepted in the popular vote. Public discussions and laws adopted in some of the cantons pointed towards public opinion in favor of clear restrictions. In addition, the "Beobachter" was a well-known consumer protection magazine disposing of a considerable mobilization power and media presence. By anticipating voter preferences and taking into account the mobilizing potential of the interests behind the initiative, Parliament decided to include the goals and prohibitions proposed by the initiative and at the same time expand them. Parliament was clearly concerned that if it stayed with the counter-proposal of the Federal Council, citizens would prefer the initiative to the counter-proposal. The choices of Parliament led in fact to the retraction of the initiative and resulted in the acceptance of the counter-proposal in 1992 with 73.8 percent support, building the foundation for further restrictive legislation.

After the constitutional article was accepted, the Federal Council proposed a restrictive Federal Law on ART in order to counteract a second popular initiative, the "Initiative for procreation respecting human dignity," which aimed at reversing the constitutional article and asked for a total prohibition of IVF and gamete donation in the federal constitution. The restrictive proposition weakened the cause of the initiative and at the same time the government did not risk an optional referendum, because the Radical Party was the only governmental party asking for a more liberal solution (i.e., one that would permit PID and egg donation) while majorities within the Christian Democrats, the Social Democrats and the People's Party wanted these techniques to be prohibited. Small parties from the Left and the Right supported the latter. Instead of the typical Center-Right or Center-Left coalition (Kriesi 2001), the decision-making process was, therefore, dominated by a rather unusual Left-Center-Right coalition, excluding the Radical Party. It gave way to a majority in Parliament supporting the prohibition of egg donation and preimplantation diagnostics. The Radical Party did not initiate a referendum in light of the support for the law by the other governmental parties and the lack of a strong partner from research or economic circles. As mentioned above, the "Initiative for procreation respecting human dignity" was defeated in the popular vote, opening the way for the federal law to take force. In the case of the Federal Law on Stem Cell Research, all governmental parties supported the solution to allow for research on leftover embryos. In the case of an optional referendum, strong support from the governmental parties considerably raises the likelihood of passing the popular vote. Given the support by the governmental parties and the more favorable attitude of Swiss citizens towards red biotechnology, a clear majority supported the Federal Law on Stem Cell Research in the popular vote.

In the case of green biotechnology, the use of the popular initiative equally influenced legislation; first, it influenced the revision of the Federal Environmental Protection Law in 1995 and the subsequent adoption of various ordinances; second, it had an effect on the elaboration of the Federal Law on Gene Technology by pushing to close existing legal gaps and by sustaining a restrictive policy design; and third, the writing of a five-year moratorium for GMOs in Swiss agriculture into the constitution.

Winning and losing parties can more easily be identified when it comes to the Gene-Protection-Initiative and the popular vote than for other parliamentary decisions. The debate around the Gene-Protection-Initiative revealed a clear Left-Right divide, with the governmental parties from the Conservative-Right, the Right and the Center (i.e., the People's Party, the Radicals and the Christian Democrats), successfully opposing the initiative and the Social Democrats unsuccessfully supporting it. The Greens, not represented in government, were also in favor of the initiative. During the elaboration of the Federal Law on Gene Technology the Right-Left divide was no longer as clear-cut. The key parties were the Radicals, the Social Democrats and the Greens. The Greens and the Social Democrats generally promoted restrictive solutions, while the Radicals asked for much more liberal policies promoting the development of this new key technology. The Christian Democrats and the People's Party were somewhat divided over certain issues. The farmer component of the People's Party promoted very restrictive policies, protecting agriculture from possible future scandals similar to the BSE scandal and other recent crises. Within the Christian Democrats, the idea of the divinity of nature led some proponents to promote restrictive policies. Interviews revealed that none of the political parties was perceived as being the clear winner in this debate. This might, however, not only have to do with consensual style of policy-making, but also with the existing legal and political situation. The Federal Law on Gene Technology built upon compromises already reached and battles won; fully restrictive solutions promoted by the Gene-Protection-Initiative had been eliminated. At the same time, a strict framework had already been established with respect to deliberative releases, contained use, and authorization, labeling and traceability of GM food products on the ground of the constitutional article and namely under the pressure of the Gene-Protection-Initiative.

During parliamentary debate, the proposition of a temporary moratorium on GMOs in Swiss agriculture failed to find a majority, but later found
the approval of the people. The popular initiative for a moratorium was supported by the Social Democrats and the Greens. The Christian Democrats,
the Liberals and the Conservatives did not officially endorse the initiative.
However, opinions within the Christian Democrats and the People's Party
were divided, and several cantonal chapters supported the initiative.
Namely the agricultural wing of the People's Party was, together with the
Farmers Association, in favor of a moratorium. Furthermore, in contrast to
the Gene-Protection-Initiative, the mobilization of industry interests
against the initiative remained limited, thus allowing the sponsors of the initiative to successfully mobilize the very critical attitudes of Swiss citizens towards green biotechnology.


## FEDERAL ADMINISTRATION: REINFORCING
## THE RESTRICTIVE WAY

Parliament played an important role in initiating legislation but also in formulating it. The "Gen-Lex" motion, which asked the Federal Council to
propose comprehensive legislation in the field of gene technology, was
voted by Parliament instead of an official, indirect counter-proposal to the
Gene-Protection-Initiative. Furthermore, the preparing commissions of
Parliament thoroughly transformed the governmental proposition of adapting different existing acts by elaborating the Federal Law on Gene Technology, regrouping the different legal provisions in one main piece of legislation, and Parliament, in comparison to the project proposed by the
Federal Council, pushed the final legislation in an even slightly more restrictive direction.[10] In fact, Parliament decided to engage its own legal expert to render its work more independent from the administration. One
should, however, keep in mind that the Federal Law on Gene Technology
was strongly based on existing norms that had been formulated on the ordinance level. Hence, the administration influenced current legislation not
only through preparing the "Gen-Lex" project, but also by adopting norms
on the ordinance level (i.e., by implementing the adapted Federal Law on
Environmental Protection and the Federal Food Law). The administration's influence was particularly strong in the implementation phase, where
its practices contributed to the overall restrictive policies. Because applications of gene technology concern various policy fields, a broad number of
administrative units and several federal departments ("ministries") were in-

volved in the elaboration and implementation of norms, namely various offices within the Federal Department of Home Affairs, the Federal Department of Economic Affairs and the Federal Department of Environment, Transport, Energy and Communication (i.e., the Federal Environmental Agency), the Swiss Federal Office for Agriculture, the Federal Veterinary Office and the Federal Office of Public Health. In addition, two expert commissions established for the agro-food sector were also influential actors (i.e., the Federal Committee for Biosecurity and the National Ethics Commission for the Non-Human Field). The most important administrative actor, however, was the Federal Environmental Agency as it had the lead in elaborating the Federal Law on Gene Technology and had considerable responsibilities in adopting ordinances for deliberative releases and contained use on the basis of the revision of the Environmental Protection Law in the mid-1990s. The Environmental Protection Agency strongly pushed the precautionary principle, while other actors adhered to more classical risk assessment procedures, namely the Federal Committee for Biosecurity, who did not share the Agency's position with respect to authorizing deliberative releases. The National Ethics Commission for the Non-Human Field seemed to position itself somewhat closer to the Federal Environmental Agency by issuing opinions on the concept of the dignity of the creature and the precautionary principle. The resulting Federal Law on Gene Technology corresponds largely to the policies promoted by the Federal Environmental Agency. Hence, the Federal Environmental Agency was able to realize its goals thanks to the support of, or rather a coalition with, the parliamentary Left and critical interest groups. This hypotheses is probably also valid if formulated the other way around, the Left and critical interest groups were able to realize some of their goals thanks to the role the Federal Environmental Agency played in preparing legislation, adopting ordinances and implementing GMO policies.

For the case of ART, the story is slightly different. On the one hand, the federal administration contributed the more restrictive solutions among the ones debated. On the other hand, diverging positions between federal departments led to postponing the regulation of embryo research and opened the door for the intermediate solution for stem cell research.

In the preparliamentary stage of adopting the Federal Law on Assisted Reproduction, two elements were particularly controversial: the prohibition of egg donation and preimplantation diagnostics. The constitutional amendment did not predefine whether these two techniques should be prohibited or not. The preferences of the members of the Federal Council and the solutions advocated by their departments differed. The Department of

Justice, headed by a Christian Democrat, and having the lead in elaborating the law, proposed to prohibit both techniques after internal debate, while the Department of Home Affairs, headed by a Social Democrat and in charge of science and research questions, wanted to allow, under certain conditions, preimplantation diagnostics and egg donation. The Federal Council as a whole decided to follow the version elaborated by the Department of Justice and sent the respective draft for a law into the preparliamentary consultation procedure. With the prohibition of egg donation and preimplantation diagnostics, those actors who would support the initiative in case the law would turn out to be too permissive could be expected to vote for the law. In addition, formulating a more restrictive solution would not risk the initiation of a referendum, given that among the governmental parties only the Radicals were opposed and in the absence of any important interest groups threatening with a referendum.

Diverging opinions were also of relevance for the question of embryo research. The Federal Law on ART does not address the question of embryo research because the task of formulating legal measures with respect to embryo research was attributed to another legislative project under the lead of the Department for Home Affairs, in particular the Federal Office for Education and Science, which instituted an expert group commission on human-subject research. In its first report, the majority of this expert commission on human-subject research recommended to allow for some embryo research under specific conditions as well as for preimplantation diagnostics to a certain extent. A minority, however, did not share this point of view, and proposed to prohibit both. The Department of Justice then produced a legal opinion on the implications of the constitutional amendment on whether to prohibit these two techniques or not. The conclusion was that the constitution did not predefine prohibition of both techniques, but that the constitutional article on ART and biotechnology overall pointed in the direction of prohibiting research on leftover embryos (Bundesamt für Justiz 1995).

Overall, it seems that the division of labor between the administrative units and their diverging interests postponed decisions on embryo research. Given the political discussions around ART, it was rather likely that strong limits would be set for any type of invasive embryo research. By attributing the task to another legislative project under the lead of the administrative department in charge of research questions, the option was kept open for adopting less limiting policies at a later date. Given the pressure of technological progress in stem cell research and the fact that the Swiss National Science Foundation authorized the funding of a research project using imported stem cells at the University of Geneva in 2003, the Federal Council decided to present a Federal Law on Embryo Research as fast as possible, which would later be integrated into the Law on Human Subject Research. Parliament preferred, however, to first address the question of embryonic stem cell research and to postpone once more from comprehensively addressing the question of embryo research, turning the project instead into a Stem Cell Research Law. In short, in the case of ART, on the one hand, the lead of the Department of Justice contributed to the restrictive framing of the issue, on the other hand, diverging opinions and the attribution of the embryo research question to the Department of Interior Affairs opened the door for an intermediate policy with respect to stem cell research.

## FEDERALISM AND COURT IMPACT: SAFEGUARDS AGAINST TOTAL PROHIBITIONS

Swiss cooperative federalism demands that the federal government and the cantons agree on the policies to be adopted in order to guarantee an effective and efficient implementation by the cantons or communes. In both fields, however, cooperative federalism was of limited importance for the elaboration of federal law. The cantons agreed on the necessity to create federal law, and where they are in charge of implementing it, the task falls within existing competences. Of relevance for the policy formulation process were, however, the important competences of the cantons and the noncentralization of the Swiss state as well as the judicialization of the two issues and the impact of the respective court decisions.

In the case of ART, the policy-making process started out at the cantonal level. The Swiss health care system is decentralized and characterized by a mixture of public and private health care providers. The cantons play a major role in formulating and implementing health policies and they are important health care providers: they are in charge of cantonal and regional hospitals, and they are notably in charge of university hospitals. As an important player in health care policies, as well as being a provider directly confronted, early on, with the questions provoked by the new techniques, some of the cantons did not want to wait for federal legislation and chose, rather, to adopt their own laws and regulations.

The design of cantonal laws and regulations varied strongly. Three cantons, Glarus (1988), St. Gallen (1988), and Basel-City (1991), prohibited almost all available ART, including full prohibitions of IVF and gamete donation. The cantonal laws of St. Gallen and Basel-City were challenged in the Swiss Federal Supreme Court, which in the case of an abstract review of a cantonal law is the court of first and final appeal. The court ruled on the first case, the canton of St. Gallen in 1989 (BGE 115 Ia 234, March 15, 1989) before the federal government published its message concerning the "Beobachterinitiative" and before the parliamentary debate took place. The court ruled that general prohibitions of certain techniques in cantonal laws were unconstitutional and questioned the practice of the anonymity of donors (BGE 115 Ia 244).

The court's decision led to policy convergence at the cantonal level, by ruling out extremely restrictive solutions. It thereby clearly influenced the starting conditions for the debate on the federal level. The arguments of the Federal Supreme Court found a strong resonance with the actors on the federal level. In particular, the opponents of total prohibitions referred to the court's opinion that general prohibitions violate the right to personal freedom. Furthermore, its jurisprudence strongly contributed to adopting the right to know one's ancestors (Rothmayr 1999; 2001).

We can also observe a judicialization of policy-making in the case of GMOs, however, only in respect to the implementation stage. The federal level attributes authorizations for deliberative releases, contained use and GM food products. The cantons play a role in implementing policies with respect to contained use, because they are in charge of developing the necessary measures for cases of major accidents. In the case of deliberative releases, the canton in which the release would take place is invited to state its opinion, yet has no veto in the matter. With respect to GM food, the cantons are in charge of controls. Overall, we could say that the cantons remained in charge of tasks with respect to food policy and major accidents in general that they had already assumed before, and the most important implementation decisions of what should be authorized with respect to GMO in the agro-food sector are taken by the federal administration. To initiate legal action against implementation decisions of the Federal Administration was one of the strategies used by interest groups critical of GM food in order to attack the policies of the Federal Office of Public Health. They filed complaints against the unauthorized use of a GM vitamin and also against the first authorization of GM soy by the Office. They also legally attacked a public campaign against the Gene-Protection-Initiative that claimed that if the initiative would be accepted, research for medical progress in the case of currently incurable illnesses would be prohibited. The judicialization[11] of the conflict was overall not a successful strategy with respect to GM food. All three complaints were rejected, and the implementation practice of the Federal Office of Public Health sustained. In the case of deliberate releases the situation was slightly different. One research institution that had been denied an authorization successfully appealed. The Federal Environmental Agency then authorized the deliberate release, but environmental groups successfully appealed this new decision. For formal legal reasons, the Federal Environmental Agency had to issue a new decision, which finally authorized the release. At the time of the adoption of the five-year moratorium, no other demands for deliberate releases were pending. Given the lack of commercial deliberate releases in Switzerland, the newly adopted moratorium does not lead to any tangible changes with respect to growing genetically modified plants in Switzerland, but rather confirms an already very restrictive authorization practice.

## LIMITED EUROPEANIZATION AND SOME POLICY TRANSFER THROUGH COMPETITION

European integration is of relevance for Swiss decision-making processes. Recent research has demonstrated (e.g., Sciarini et al. 2002; Fischer 2005) how European integration can considerably change opportunity structures for national actors. For the two fields studied, European policies had some agenda-setting effects, but limited impact on the content of national norms. In the case of GMOs in the agro-food sector, in the "Eurolex package" prepared for the Swiss vote on becoming a member of the European Common Market in 1992, a minimal gene technology regulation was proposed for the first time. After the rejection of participating in the European Common Market the proposition was reintegrated into the revision of the Environmental Protection Law. Besides this agenda-setting effect, the compatibility with European norms remained of some, but still rather limited, importance for Swiss legislation. It would go too far to say that the rather strict EU policies and the discussions leading to them, created new or better opportunities for critical interest groups because the European Union's move to more restrictive policies did not precede the Swiss shift starting in the mid-1990s and because direct democracy offered ample venues for critical interest groups to pressure for restrictive solutions. The fact, however, that the EU did not adopt policies going against the Swiss restrictive regulations was certainly of advantage to critical interest groups.

In the case of ART, Swiss regulation goes far beyond the content of the Bioethics convention and was adopted before the convention. Furthermore, there are no indications of a policy transfer, even though policies adopted by other countries were also taken into consideration. There are references at least to the following countries: Germany, Sweden, the Netherlands, Great Britain, Austria, Denmark, France, Norway, and Italy. Reports of the Council of Europe and other documents elaborated on the supranational level were taken into account. The policies adopted by other countries mainly served to position the designing solutions discussed in Switzerland within the European and international context. Among the different countries listed, it seems that the German embryo protection law and the German discussion, and also the French design and discussion, played the most prominent role. The final design clearly leans to the more restrictive German policies than the more permissive French ones. Media analysis and opinion polls indicate considerable cultural differences between the German-speaking part of Switzerland, and the French- and Italian-speaking parts: Swiss-Germans were clearly more critical towards ART and biotechnology in general (Bonfadelli et al. 1998: 154; Bonfadelli et al. 2001; Maeder 1992).

In terms of international economic competition, there seem to be considerable differences between the two fields. For ART, competition was of little relevance. Switzerland is not a forerunner country in this field and since early on, even before the adoption of the constitutional article, ART tourism to other countries with more liberal practices had already developed. For stem cell research, international competition was more important and contributed to Swiss researchers and the NSF making the first step in importing embryonic stem cell lines and applying pressure for a solution allowing stem cell research in Switzerland. While the Swiss agro-food sector is a small market of limited economic importance, the pharmaceutical and biotechnology industries are economically important. International economic competition with respect to research and development led to support for the adoption of an intermediate solution for embryonic stem cell research and was a mobilizing factor in the successful campaign of research and industry against the Gene-Protection-Initiative. With respect to economic competition, the result of the trade conflict opposing the US/Canada and the EU is also of interest for Switzerland, whose regulation is similar to the EU one. Even though Switzerland is not a party in this conflict, the result will either support the existing Swiss regulation or contribute to initiating changes into a less restrictive direction.

## CONCLUSION

Switzerland adopted fairly restrictive policies for governing GMOs in the agro-food sector as well as for ART, with the exception of embryonic stem cell research where Switzerland adopted intermediate policies, and hence chose a different solution compared to other countries with similarly restrictive policies for applying ART such as Germany and Austria. While the designing processes developed differently in the two fields, they share a number of characteristics explaining the similar outcomes (see table 11.3).

There are first of all considerable parallels with respect to Europeanization and policy transfer through competition. While the international context cannot be ignored, it nevertheless seems that internationalization and European integration are of some but, compared to other explanatory factors, still limited importance for explaining the content of Swiss policies. In the field of ART, the scope of supranational norms is not only limited, but Switzerland adopted the main piece of national regulation, the Federal Law on Assisted Procreation, before the convention was signed. As discussed, EU regulation for GMOs in the agro-food sector was also not a determining factor for Switzerland choosing a restrictive design. If, however, EU regulations had turned out to be much less strict, this would have likely influenced the Swiss debate. In the case of embryonic stem cell research, policy transfer through competition played a role. Switzerland has a strong

**Table 11.3. Comparison**

| Hypotheses | ART | GMO |
|---|---|---|
| Policy Network | | |
| H1: Policy community of targets | NO | NO |
| H2: Broad mobilization by critical interest groups in issue network | YES | YES |
| | | |
| Country Pattern | | |
| H3: Christian Democrats in government/Green Party key player | YES | YES |
| H4: Critical groups pushing on national and federated level (fragmented governance with coordination) | YES | YES |
| H5: Ministries in charge not sharing interests of target groups | YES/NO* | YES |
| | | |
| Internationalization | | |
| H6: Europeanization/European policies successfully used by critical interest groups | NO | NO |
| H7: Policy transfer through competition | NO/YES* | NO |

* for embryonic stem cell research

economic interest in research and development in the sector of red biotechnology. International research competition together with other factors,
namely the distribution of regulatory tasks among federal offices and the
strategically motivated postponing of comprehensively regulating research
on leftover embryos, helps to explain why Switzerland chose an intermediate and not a restrictive design for stem cell research.

Overall, in comparison to international factors, mobilization of critical interest groups, their use of direct democracy, the influence of the federal departments and offices in charge of preparing legislation, and party politics
are more helpful in explaining the similar results of the two policy-making
processes. First of all, we find a strong mobilization of critical interest
groups in both fields, probably even a stronger one with respect to biotechnology in general than specifically for ART. Critical interest groups successfully influenced the content into restrictive directions through the use
of the instrument of popular initiatives. The first popular initiative, the
Beobachterinitiative, was crucial for the "problematization" of the issue of
biotechnology and framed the debate in terms of the need to protect human beings and nature from the possible negative effects of this new technology, a frame that dominated and continues to dominate the debate
within legislative processes. The most recent initiative instituted a five-year
moratorium for a GMO-free Swiss agriculture, while the other initiatives
helped to push legislation into a restrictive direction in general. Critical interest groups also successfully influenced the content in both fields through
their lobbying and privileged ties—or at least similar interests—with some
of the administrative units in charge of preparing legislation or implementing public policies, in the case of green biotechnology namely the Federal
Environmental Agency. In the case of ART, the usual health policy networks were not of relevance, because the Department of Justice was in
charge of elaborating the Federal Law on Assisted Reproduction. This also
contributed to explaining the lack of influence of the self-regulatory regime
of the medical community in the case of ART. Furthermore, among the
governmental parties, in both fields, only the Liberals promoted clearly less
restrictive policies. In the case of ART, a large majority of the Christian
Democrats, the Social Democrats and the People's Party, under the pressure of the popular initiative, chose restrictive solutions over intermediate
policies. For embryonic stem cells, more specifically, the situation was different insofar as we found a broad consensus among all four governmental
parties. With respect to the Gene-Protection-Initiative, only the Social
Democrats among the governmental parties supported the initiative. When
it came to legislating on the Federal Law on Gene Technology, however,

not only the Social Democrats promoted restrictive solutions, but also the Christian Democrats, the agricultural part of the Conservatives, and the Greens, as the biggest nongovernmental party. The broad consensus for solutions that were restrictive, but against total prohibitions, was also the result of the successful campaigning and lobbying of industry and researchers, who strongly mobilized around the Gene-Protection-Initiative. In addition, federal structures and judicialization helped to counteract attempts to fully prohibit basic techniques. In the case of ART, the intervention of the Swiss Federal Supreme Court contributed to turning cantonal policies in a less restrictive direction and indirectly influenced the debate on the federal level by declaring total prohibitions of basic techniques to be violating the right to personal freedom. Because of the limited power of the Swiss Federal Supreme Court to review federal legislation, in the case of GMOs, the court cases only concerned implementation decisions on the authorization of field experiments and food containing GMOs. By deferring to administrative practices, in the case of food, and by criticizing the restrictive authorization practice for deliberate releases, court instances contributed to preventing de facto prohibitions.

Regulation in both fields is still evolving. Embryo research still needs to be addressed in a comprehensive manner through the pending legislation on human subject research. Attempts, however, to change ART legislation into a less restrictive direction (e.g., by allowing for preimplantation diagnostics) have all failed. For GMOs in the agro-food sector, it remains to be seen what will happen once the moratorium ends and what impact the trade conflict opposing the US and Canada against the EU might have for further developments in Switzerland.

## NOTES

1. I would like to thank Uwe Serdült and Michel Berclaz, who contributed to the data collection at various points in the project. My thanks also go to the Swiss National Science Foundation, who contributed to financing a part of this research.

2. Laws and decrees of the Federal Assembly can be challenged through the *optional referendum*, which requires fifty thousand signatures of citizens. The referendum has to be understood as an intervention, or the possibility of a veto, at the end of the policy formulation and decision-making process (Kriesi 1998). Its importance lies not only in the acceptance or rejection of the result of the decision-making process, but in its latent impacts on the policy-formulation process as a whole (see Linder 1994: 118f. Papadopulos 2001). The outcome of a referendum is always uncertain and the success rate in popular ballots at 60 percent is fairly high

(Linder 1994: 100). Therefore, political actors try to negotiate a compromise that satisfies the interests that have the potential to initiate and win a referendum.

3. See also ordinances FrsV SR 814.911 and ESV 814.912 (August 25, 1999), both under revision by February 2006 in order to account for the new gene protection law, the Cartagena Protocol, decisions of the Swiss Federal Supreme Court, and the latest version of EU regulations.

4. See 'Lebensmittelgesetz', 'Lebensmittelverordnung' und 'Verordnung über das Bewilligungsverfahren für GVO-Lebensmittel, GVO-Zusatzstroffe und GVO-Verabeitungsstoffe (VGBO).'

5. The cryopreservation of impregnated eggs is permitted.

6. Swiss norms also allow for "negative" labeling (i.e., using the fact that a food product does not contain GMO for marketing purposes).

7. Swiss legislation also covers uses of GMO in fertilizer and pesticides and establishes for feed authorization, labeling and traceability requirements.

8. The Federal Data Protection Commission, the Federal Office of Civil Law Affairs and the Federal Office of Statistics are also involved in the implementation.

9. The two main umbrella organizations are: Interpharma (Roche, Novartis, Serono) and Internutrition (Föderation der Schweiz, Nahrungsmittel-Industrien, Hoffmann-La Roche AG, Monsanto, Nestlé Suisse SA, Syngenta, Novozymes, Wander AG, Du Pont).

10. This was only partly manageable as several other legislative acts still needed to be changed, along with the adoption of the Federal Law on Gene Technology.

11. The tradition of supremacy of Parliament is still very strong in Switzerland. The Swiss Federal Supreme Court has no power to overturn federal laws (i.e., its power of nullification is limited).

## REFERENCES

Bonfadelli, H. 2000. Gentechnologie im Urteil der Bevölkerung: Agenda-Setting–Wissensklüfte–Konsonanzeffekte. In *Gentechnologie in der öffentlichen Kontroverse*. Eds. H. Bonfadelli and U. Dahinden. Pp. 46–96. Zürich: Seismo.

Bonfadelli, H., P. Hieber, et al. 1998. Switzerland. In *Biotechnology in the Public Sphere: A European Sourcebook*. Eds. J. Durant, M. W. Bauer and G. Gaskell. Pp. 144–61. London: Science Museum.

Bonfadelli, H., and U. Dahinden (eds). 2000. *Gentechnologie in der öffentlichen Kontroverse*. Zürich: Seismo.

Bonfadelli, H., U. Dahinden, et al. 2001. Biotechnology in Switzerland: From Street Demonstrations to Regulations. In *Biotechnology 1996–2000: The Years of Controversy*. Eds. G. Gaskell and M. W. Bauer. Pp. 282–91. London: Science Museum.

Bundesamt für Justiz. 1995. *Art: 24novies BV 'Fortpflanzungsmedizin, Verfassungsrechtlicher Status von Embryonen'*, VPB 60.67, 17. November.

Dahinden, U. 2000. Die Schweizer Gentechnologie-Debatte im internationalen Vergleich. In *Gentechnologie in der öffentlichen Kontroverse*. Eds. H. Bonfadelli and U. Dahinden. Pp. 191–203. Zürich: Seismo.

Eisner, M., N. Graf, and P. Moser (eds). 2003. *Risikodiskurse. Die Dynamik öffentlicher Debatten über Umwelt- und Risikoprobleme in der Schweiz*. Zürich: Seismo.

Fischer, A. 2005. *Die Auswirkungen der Internationalisierung und Europäisierung auf Schweizer Entscheidungsprozesse. Institutionen, Kräfteverhältnisse und Akteursstrategien in Bewegung*. Chur: Rüegger.

GfS-Forschungsinstitut. 2003. *Klare Präferenzen bei der Anwendung. Schlussbericht zum Gentechnik-Monitor 2003 für die Interpharma*. Bern: GfS.

Graf, N. 2003. Die Last von 30 Jahren Oekologiediskurs: Alte und neue Deutungsmuster in der Gentechnolgiedebatte. In *Risikodiskurse. Die Dynamik öffentlicher Debatten über Umwelt- und Risikoprobleme in der Schweiz*. Eds. M. Eisner, N. Graf, and P. Moser. Pp. 212–40. Zürich: Seismo.

Kriesi, H. 1998. *Le système politique suisse*. Paris: Economica.

Kriesi, H. 2001. The Federal Parliament: The Limits of Institutional Reform. In *The Swiss Labyrinth, Institutions, Outcomes and Redesign*. Ed. J.-E. Lane. London: Frank Cass.

Linder, W. 1994. *Swiss Democracy: Possible Solutions to Conflict in Multicultural Societies*. Houndsmill: Macmillan.

Maeder, C. 1992. Reproduktionsmedizin in der Schweiz: Ergebnisse und Interpretationen einer repräsentativen Bevölkerungsbefragung. *Schweizerische Zeitschrift für Soziologie* 18: 363–91.

Moroni, I. 1994. Processus de politisation des problèmes et mouvements féministes: le cas de l'avortement et de la procéation assistée en Suisse. In *Femmes et politiques*. Annuaire ASSP. Pp. 99–122. Bern: Haupt.

Papadopoulos, Y. 2001. How Does Direct Democracy Matter? The Impact of Referendum Votes on Politics and Policy-Making. In *The Swiss Labyrinth, Institutions, Outcomes and Redesign*. Ed. J.-E. Lane. London: Frank Cass.

Rothmayr, C. 2003. Politikformulierung in der Fortpflanzungstechnologie: Partizipation und Einfluss feministischer Gruppierungen im internationalen Vergleich. *Österreichische Zeitschrift für Politikwissenschaft 2003* (2): 189–200.

———. 2001. Towards the Judicialisation of Swiss Politics? In *The Swiss Labyrinth, Institutions, Outcomes and Redesign*. Ed. J.-E. Lane. London: Frank Cass.

———. 1999. *Politik vor Gericht*. Bern: Haupt.

Schweizerische Akademie der Medizinischen Wissenschaften (SAMW). 1981. Medizinisch-ethische Richtlinien für die artifizielle Insemination. Basel: SAMW.

———. 1985. Medizinisch-ethische Richtlinien zu In-vitro-Fertilisation und Embryotransfer. Basel: SAMW.

———. 1990. Medizinisch-ethische Richtlinien für die ärztlich assistierte Fortpflanzung. Basel: SAMW.

Sciarini, P., et al. 2002. L'impact de l'internationalisation sur les processus de décision en Suisse. *Revue suisse de science politique* 8 (3/4): 1–33.

# ⑫

# REGULATING ART AND GMOS IN EUROPE AND NORTH AMERICA: A QUALITATIVE COMPARATIVE ANALYSIS

*Éric Montpetit (Université de Montréal),*
*Frédéric Varone (Université de Genève), and*
*Christine Rothmayr (Université de Montréal)*

In the first chapter, we presented three common approaches to explain public policies. First, the policy network approach helps to explain variations in policy choices across sectors by focusing on sectoral policy networks. Our use of this approach has shown that significant policy variations may exist between GMO and ART policies within a single country, depending on the presence of a policy community of potential target groups, or on the mobilization of actors concerned about biotechnology in an issue network. Secondly, the country pattern approach is useful to explain the likelihood of consistent policy choices between ART and GMO within each country, but can also be used to predict policy variations between countries based on institutional differences, such as governing parties, political institutions and administrative organizations. Thirdly, the internationalization approach can provide an explanation for policy similarities across countries and sectors. The mobilization of international rules by target groups or the initiation of policy transfer might result in comparably restrictive or permissive policies. However, we did not predict that international rules and policy transfers would lead to overall policy convergence across all countries and both policy sectors, as the interests of actors successfully mobilizing international rules and initiating policy transfer might differ across countries and sectors.

Each of the authors of the country chapters was asked to analyze the impacts on the policy design of the variables falling under each of these three approaches. The in-depth qualitative analyses conducted in each of the countries enabled the authors to trace the influence of the seven variables on the policy design. The following comparison of countries and sectors builds on the case studies' findings. The comparison allows us to address the hypotheses formulated in chapter 1 and, thus, to measure the contribution of each of the three approaches to explaining policies for ART and GMOs. In order to conduct a systematic comparison, the authors of individual country chapters filled out a table identifying the extent to which the factors falling under the three approaches contributed to the policy choices in the ART and GMO sectors in their respective country. The Qualitative Comparative Analysis,[1] described below, is based on these tables produced by the country specialists. The Qualitative Comparative Analysis emphasizes the idea that each of the three approaches, the policy network, the country pattern and the internationalization approach, contributes some elements to explaining biotechnology policy choices. It also enables us to identify specific combinations of conditions whereby policy-makers come to adopt permissive, intermediate or restrictive biotechnology policies. The comparison rests on the idea that combinations of conditions are more apt to explaining policy choices than variables taken individually. In this respect, our analysis departs from existing biotechnology policy studies, which often propose one approach that focuses on one or two explanatory variables for explaining policies. The following section describes the Qualitative Comparative Analysis's method in more detail.

## QUALITATIVE COMPARATIVE ANALYSIS IN A NUTSHELL

Our comparative analysis is based on the Qualitative Comparative Analysis (QCA), which seeks to identify patterns of multiple conjunctural causation and is particularly well suited for "small-N" research designs (Ragin 1987; De Meur and Rihoux 2002). The purpose of the QCA method is to "integrate the best features of the case-oriented approach with the best features of the variable-oriented approach" (Ragin 1987: 84). In fact, the QCA possesses some of the key strengths of qualitative research, namely seeking holistic and case-sensitive qualities (De Meur and Rihoux 2002: 20f.). The method considers that each individual case can require a specific explanation. The method thus employs a complex conception of causality. First, independent variables, or "conditions" in QCA terminology, combine in most cases to produce any particular policy output. Second, different combinations of conditions can produce the same policy output. Third, in a different combination, a given condition can contribute to contrasting policy outputs. When this occurs, the combination of conditions is more important than the individual conditions in the explanation of the policy output. This complex understanding of causality is captured by the concept of "multiple conjunctural causation" in QCA terminology. Using QCA, the researcher is urged not to "specify a single causal model that fits the data best," but instead to "determine the number and character of the different causal models that exist among comparable cases" (Ragin 1987: 167). In other words, we use the QCA to identify the different conjunctures of conditions leading to restrictive or permissive biotechnology policy in different countries.

The QCA also shares some of the key strengths of quantitative methods (De Meur and Rihoux 2002: 20f.). The QCA allows for the analysis of more than simply a few cases and can thereby produce limited generalizations. Moreover, it relies on formal tools, namely Boolean algebra, and allows discussions of individual variables, conceived as conditions, in the explanation of policy outputs. QCAs are also replicable, enabling other researchers to corroborate or to falsify results. Lastly, computer-generated Boolean algebra allows one to identify "causal regularities" that are parsimonious, combining only a few minimum conditions.

The process of the QCA begins with the production of raw data tables in which cases are summarized as combinations of conditions, expressed in 0 or 1 values, related to a policy output, also expressed in 0 or 1 values. As mentioned above, the authors of each of the individual country chapters provided the raw data tables used in the comparison. From the raw data, we used software to generate a "truth table" displaying the configuration of conditions that led to the policy outputs in each and every country and sector. The truth table revealed, in a rough manner, the extent to which similar policy outputs across several countries resulted from similar or different configurations of conditions. The next step of the analysis was Boolean minimization. Here the QCA software reduced the long Boolean expression, which is a full description of the truth table, into shorter expressions, the so-called "minimal equations." The minimal equations not only indicate whether the same policy output results from the same combination of conditions, it accounts for the presence or absence of given conditions in the production of every single policy output. In other words, the minimal equations reveal causal regularities, in the form of combinations of conditions that most influenced the policy output. It should be noted that in some cases, two or more minimal equations could potentially fit the raw data of a country. When this happens, the researcher must decide which minimal equation makes the most sense, given his or her knowledge of the case. This situation only presented itself in the cases of Canada and the United States, two countries for which the authors of this comparative chapter have in-depth qualitative knowledge.

The QCA, we show in the rest of this chapter, was invaluable in helping us identify patterns of biotechnology policy-making across several countries without neglecting the full complexity of policy-making within each and every country. The result may not support a single, cohesive and elegant theoretical model, but it provides a realistic account of policy-making in a variety of contexts by drawing from a plurality of theories. One of this book's significant contributions is to show that a general policy explanation, draw-ing from a narrow theoretical source, risks missing what really matters for policy-making in specific contexts. Meanwhile, capturing the richness of a policy-making context does not have to imply giving up on regularities that extend across countries and sectors. Put simply, capturing the richness of a policy-making context requires taking some distance from basic under-standings of causation, acknowledging patterns of multiple conjunctural causation, which is the case with QCA.

Before moving on to the results of the QCA, it would be useful to discuss the observed variations in terms of the restrictiveness and permissiveness between ART and GMO policies across countries. This discussion of the variation will allow us to draw preliminary conclusions on the plausibility of networks, country patterns and internationalization as explanations for biotechnology policy.

## BIOTECHNOLOGY POLICY ACROSS COUNTRIES AND SECTORS

The authors of the country chapters classified their country's ART and GMO policies into three categories: permissive, restrictive or intermediate (for details see chapter 1). Table 12.1 summarizes and compares the find-ings for GMO and ART policies in the nine countries.

As we explained in the introduction of this chapter, we drew independent variables from three common approaches in comparative policy studies. These three approaches are based on different understandings of policy-making. The first (i.e., policy network approach) adopts a bottom-up per-spective on policy-making whereby each policy sector displays its own con-straints and opportunities. This approach stresses the nature of policy sectors as key to understanding policy choices. For example, resourceful multinational firms are far more present in the GMO sector than in the

**Table 12.1. Biotechnology Policy in Nine Countries and Two Sectors**

| Agro-food genetic engineering (GMOs) | Assisted Reproductive Technology (ART) | | | Number of empirical observations for GMOs in each category |
| --- | --- | --- | --- | --- |
| | Restrictive | Intermediate | Permissive | |
| Restrictive | Switzerland (CH), Germany (G) | Sweden (SW) | United Kingdom (UK) | 4 |
| Intermediate | | France (F) The Netherlands (NL) | The Belgium (B) | 3 |
| Permissive | | Canada (C) | United States of America (USA) | 2 |
| Number of empirical observations for ART in each category | 2 | 4 | 3 | |

ART sector. Moreover, the ART sector raises moral concerns and serious questions about the character of human life, which should have an impact on the mobilization of biotechnology opponents distinct from that of the environmental uncertainties related to GMOs. In short, the nature of each of the two sectors should encourage the formation of distinctive policy networks likely to press for the adoption of distinctive policies. A basic understanding of the relationship between networks and policy would therefore predict differences between ART and GMO policies but consistency across countries. Table 12.1 fails to provide support for this basic understanding of the effect of policy networks. In fact, permissive, intermediate and restrictive policies were observed in each of the two sectors; ART policies are inconsistent across the nine countries as are GMO policies. Moreover, ART and GMO policies are consistent in five of the nine countries. In fact, Switzerland, Germany, France, the Netherlands and the United States have restrictive, intermediate or permissive policies in both ART and GMO sectors.

The second (i.e., the country pattern approach) suggests that it is not so much the nature of policy sectors that provides policy-makers with constraints and opportunities when making policy choices as it is the nature of political institutions. In some countries, notably federations, institutions fragment policy processes, thus providing different constraints and opportunities to biotechnology opponents and proponents. In other countries, power is concentrated in the hands of a few elected officials who can make decisions across sectors that are consistent with their values and beliefs. A basic understanding of the country pattern approach would therefore predict different policy choices between countries whose institutions, on the one hand, fragment governance and countries whose institutions, on the other hand, concentrate governance in the hands of few policy-makers. However, the country pattern approach would predict great consistency between ART and GMO policies within each and every country. Again, table 12.1 fails to provide support to this basic understanding of the country pattern approach. Four out of the nine countries made inconsistent policy choices between ART and GMOs. Moreover, Switzerland, the United States and Canada, arguably the three countries whose institutions fragment governance the most, are widely spread out in the table, indicating that they have made sharply different policy choices. Sweden, the United Kingdom and France, the three countries whose institutions concentrate governance the most, may not be as spread out as federal countries, but they have adopted different policies and are thus in different boxes in table 12.1.

The third (i.e., the internationalization approach) encourages a top-down perspective on policy-making in which the opportunities for and constraints on policy-makers would not so much come from their country's institutions as they would come from international or foreign norms and policies. From this perspective, policy-makers would draw from foreign experiences, circulated by international organizations that focus on promoting best practices. Alternatively, policy-makers could be fully constrained by international regulations into adopting a given policy. Such a basic understanding of the impact of internationalization on public policy would predict policy convergence among countries. However, table 12.1 provides no support for this prediction. In fact, the findings indicate that countries have made sharply divergent biotechnology policy choices. Overall, the results displayed in table 12.1 support the idea that explanations for biotechnology policies cannot rest on a single theoretical approach to policy-making and thereby invite us to endorse a complex understanding of causality.

As explained in the previous section, we employ the QCA method to identify multiple causal conjunctures. The comparison of policy choices (see table 12.1), however, already points to some preliminary conclusions. First, ART is less often restricted by policy than GMOs. Indeed, four countries have adopted restrictive GMO policies and only two have similarly restrictive ART policies. Conversely, only five countries have either permissive or intermediate GMO policies while seven have similar ART policies. In other words, the nature of the policy sector may have had some impact on biotechnology policies. Second, the two North American countries, table 12.1 reveals, have more permissive biotechnology policies than European countries. When combined with the observation that GMO policies tend to be more restrictive than ART policies, the observation is consistent with the idea that regional integration has a policy impact. In fact, it is now common knowledge that the European Union has adopted strict regulations on GMOs. Meanwhile, the North American Free Trade Agreement has no provision specific to biotechnology. In other words, table 12.1 suggests treating seriously the hypotheses presented under the internationalization approach. Lastly, five out of the nine countries made consistent policy choices between ART and GMO, including Switzerland and Germany, two Germanic countries. Although Sweden did not make consistent choices, like Switzerland and Germany, it leans towards the restrictive side of biotechnology policy choices. All these observations suggest that the country pattern approach should not be discredited too rapidly. Indeed, Germanic and Nordic countries share some institutional and cultural traits and were, to various extents, marked by the experiments on humans conducted under the Third Reich.

This analysis of table 12.1 is reassuring insofar as it lends support to the notion that the three approaches can contribute something to the explanation of biotechnology choices, to the extent that one rejects simplistic understandings of causation. While none of the three policy approaches suffices to understand biotechnology policy choices, it appears likely that the conditions associated with each of the approaches combine in different ways to influence policy choices. In the next section, we identify the various combinations, or multiple causal conjunctures with the help of the QCA.

## THE QUALITATIVE COMPARATIVE ANALYSIS

Regrouped under the policy network, country pattern and internationalization approaches are seven hypotheses. Each was discussed at length in the introductory chapter (for coding see Appendix 12.1). In the individual country chapters, there are findings supportive of each of the hypotheses. A policy community of targets, for example, played a significant role in making the American GMO policy permissive. The United Kingdom's restrictive GMO policy is partly explained by the mobilization of opponents into issue networks. A governing party with conservative values played a role in the development of the relatively restrictive French ART policy. The fragmented system of power separation in the United States contributed to the country's permissive ART policy. In Germany, administrative preferences coincided with those of opponents to biotechnology, exerting pressure for the adoption of restrictive policies in both the ART and the GMO sectors. In every single member state of the European Union, opponents successfully used European policy to prevent the adoption of permissive GMO policies. And in Switzerland, target groups in the case of embryonic stem cell research successfully initiated a policy transfer.

These are only illustrations that are of little help in sorting out idiosyncratic phenomenon from relatively general causal factors or in assessing the relative weight of conditions contributing to a policy explanation. The purpose of this final chapter is to depart from such illustrations. The QCA enables us to identify the minimal combinations of conditions necessary for the adoption of permissive, intermediate or restrictive biotechnology policies. As explained above, the QCA begins with the construction of a so-called "truth table" (see Appendix 12.2). The purpose of the table is to organize the empirical observations in a manner enabling the identification of configurations of conditions leading to every possible policy outputs. The truth table presents the combination of conditions prevailing in each country, according to the restrictive, intermediate or permissive character of their ART and GMO policies.

The truth table reveals that each case is not entirely unique. The same combination of conditions led Germany in the ART sector and Switzerland in both ART and GMOs to adopt restrictive policies. Canada and the United States' permissive GMO policies are also explained by the same combination of conditions. And lastly, the restrictive GMO policies of the UK and Sweden also coincide with the same combination. More importantly, the truth table displays no contradiction; we found no given combination of conditions associated with different outputs. In other words, cases sharing a combination of conditions are never associated with contrasting policy outputs. This is an indication that the conditions identified in our hypotheses exert strong discriminating explanatory power and that the conditions we selected for this study are appropriate.

This being said, we ended up with thirteen conditions for eighteen cases, potentially creating a degree of freedom problem. A degree of freedom problem occurs when the number of independent variables, or conditions, comes too close to the number of cases. A degree of freedom problem can be corrected by eliminating the least useful variables or conditions from the analysis. De Meur and Berg-Schlosser (1996) suggest a method of variable elimination inspired by Przeworski and Teune's (1970) "most different and most similar system" of comparison. The method is premised on the idea that a variable cutting across most different systems, irrespective of variations in policy outputs, has no explanatory power and therefore can be eliminated from the analysis. Likewise, a constant between similar cases whose policy outputs sharply contrast also has no explanatory power and can be removed from the analysis. While we do not formally use the distinction between most similar and most different systems, we can nevertheless rely on the idea that a constant condition across varying policy outputs has no causal relevance and can therefore be eliminated from the analysis. As indicated by the truth table, the absence of international rules used by targets (IRT), the absence of policy transfer initiated by targets (PTT) and the absence of policy transfer initiated by opponents (pto) are constant conditions between restrictive, intermediate and permissive cases. Once these three conditions are eliminated, we are left with ten conditions, relieving us from our degree of freedom problem.

From the truth table data, we conducted a Boolean minimization through pair wise comparisons. The QCA computer software (TSOMANA 1.25, Cronqvist 2006) eliminates the most irrelevant conditions, which could not be seen from a visual examination of the truth table. The purpose of this operation is to reduce the complexity of the combinations of conditions associated with each policy output. The procedure identifies the so-called "prime implicants," which form the most concise possible combinations of conditions associated with each possible policy output. These combinations are presented in the form of Boolean equations. Conditions in uppercase letters are those that received value 1 in the truth table and conditions in lowercase received value 0. Naturally, combinations or equations are associated with policy outputs. The factorized minimal equations, in which we have emphasized common conditions, are presented in table 12.2.

Table12.2 reveals that countries with restrictive policies share three factors. Restrictive biotechnology policies in the ART and the GMO sectors were adopted where opponents and actors concerned with biotechnology joined efforts in an issue network, where administrative preferences matched the preferences of groups opposing or concerned about biotechnology and where governance was not concentrated. In the case of GMOs, the use of international rules—EU regulations—by opponents of biotechnology contributed to the restrictive output. At the same time, equation 1 indicates that the presence of a conservative government, or control of key ministerial positions by parties with restrictive preferences, is not an essential condition for adopting restrictive policies. Nevertheless, Switzerland (ART and GMO) and Germany (ART) provide evidence that the presence of Christian Democrats in government can contribute to the design of restrictive biotechnology policies, although Left-leaning parties in these two countries also had a preference for restrictive policies in both the ART and GMO sectors. It is also worth underlining that in the ART and GMO sectors in Switzerland, the actors that were concerned or opposed to biotechnology did not use international or European rules to obtain restrictive policies, mostly due to bad timing. The Swiss restrictive biotechnology policies are thus explained first and foremost by domestic factors.

These findings are partly consistent with existing analyses of biotechnology policy development. Toke and Marsh (2003) argue that issue networks and administrative preferences can contribute to increasing the restrictiveness of biotechnology policy. Our comparison confirms their results. Bernauer and Meins's (2003) institutional analysis suggests that multiple points of access to policy-makers help actors who are pressing for policy restrictions, which corresponds to our finding that concentrated governance prevents the adoption of restrictive biotechnology policies. Likewise, Young's (2003) argument that international rules contribute to increased restrictiveness, an argument that appears counterintuitive in light of the glob-

**Table 12.2.  Minimal Boolean Equations**

| Policy design type | Minimal equations | | Equation Number | Countries covered by the minimal equations for the ART sector | Countries covered by the minimal equations for the GMO sector |
|---|---|---|---|---|---|
| Restrictive biotechnology policies (6 cases) | INO * APO * COG | cpl * IRO | 1 | | Germany, United Kingdom and Sweden |
| | | CPL * iro | 2 | Switzerland and Germany | Switzerland |
| Intermediate biotechnology policies (7 cases) | INO * COG * | CPL * APT | 3 | | Belgium |
| | IRO * PCT | cpl * apt | 4 | | Netherlands |
| | ino * cog * iro | PCT * cpl * APT | 5 | Sweden | |
| | | pct * CPL * apt | 6 | Netherlands | |
| | INO*APT | COG * pct * cpl | 7 | Canada | France |
| | | cog * iro * PCT * CPL | 8 | France | |
| Permissive biotechnology policies (5 cases) | ino * apo * iro | cpl * COG | 9 | United States | United States and Canada |
| | | cpl * pct | 10 | United Kingdom | |
| | | COG * pct | 11 | Belgium | |

Abbreviations (for details see Appendix 12.1):
APO: Administrative preferences jibing with opponents / apo: Absence of administrative preferences jibing with opponents
APT: Absence of administrative preferences jibing with targets / apt: Administrative preferences jibing with targets
COG: Absence of concentrated governance / cog: Concentrated governance
CPL: Conservative party label / cpl: Absence of conservative party label
INO: Issue network of opponents / ino: Absence of an issue network of opponents
IRO: International rules used by opponents / iro: Absence of international rules used by opponents
PCT: Absence of a policy community of targets / pct: Policy community of targets

alization discourse, matches our observations in the GMO area. In fact, as Vogel (2004) argues, we can observe a shift in EU regulatory policies on food safety over the last decades towards the application of the precautionary principle and risk adverse policies.

Four countries have adopted permissive policies in one or two of the sectors: the UK and Belgium in the ART sector, Canada in the GMO sector, and the United States in both sectors. These five permissive cases have three factors in common: an absence of issue networks, an absence of administrative preferences matching those of opponents, and an absence of international rules used by opponents. Three additional conditions combined in different manners over the five cases to explain permissive designs (equations 9 to 11): an absence of a conservative party, an absence of concentrated governance and the presence of a policy community. In the United States' ART sector (equation 9), the absence of a dominating conservative party and the absence of concentrated governance, in addition to the three common conditions, combined to enable a permissive policy. At first sight, a combination of absent conditions to explain a policy output may surprise the reader. As chapter 3 makes clear, however, the country's permissive ART policy rests on nondecisions (Bacharch and Baratz 1963). In other words, the combination of absent conditions in the American ART sector explains the absence of federal ART decisions. In contrast, the presence of a policy community of target groups explains the adoption of permissive GMO policies in the United States. Equations 9 and 10 fully capture this difference in the United States between nondecisions over ART and permissive GMO decisions. In fact, the presence of policy communities of targets is part of the explanation of the four other permissive cases: ART policies in Belgium and the United Kingdom and GMO policies in Canada and the United States. The role of closed policy networks, which we refer to as policy communities, in the production of permissive policy outputs has already been highlighted in other research in the area of biotechnology (Montpetit 2005; Bleiklie, Goggin and Rothmayr 2004; Varone, Rothmayr, and Montpetit 2006; Toke and Marsh 2003).

The comparison of permissive and restrictive cases underlines the importance of the combination of network conditions with administrative preferences in explanations of ART and GMO policies. Restrictive policies were always adopted when issue networks of opponents were present, as was the case in Switzerland and Germany for both the ART and the GMO sectors, as well as in the UK and Swedish GMO sectors. In contrast, permissive policies were adopted where these issue networks were absent: the American, British and Belgium ART sector, as well as the Canadian and

American GMO sector. Likewise, restrictive policies were adopted when administrative preferences matched those of actors opposing or concerned about biotechnology and permissive policies when administrative preferences were dissimilar to those of these latter actors.

Furthermore, the comparison indicates that several conditions were present in restrictive and permissive cases alike. The three restrictive GMO cases, Germany, the United Kingdom and Switzerland, share with the United States and Canada, the two permissive GMO cases, the absence of a conservative party in government. Likewise, the absence of concentrated governance is a condition that Swiss and German restrictive ART policies have in common with the American and Belgium permissive ART design. In addition, the absence of international rules used by opponents is observed in the ART and GMO sectors for both restrictive and permissive cases. While conventional comparative policy analyses might treat such situations as particularly puzzling, the QCA allows for a single condition to play out differently, depending on the combination of conditions in which it is inserted. For example, the absence of concentrated governance contributed to the adoption of a permissive ART policy in the United States because it combined with the absence of issue networks, of international rules and of administrative preferences favorable to opponents. In such a case, the logic of veto points, explicitly related by George Tsebelis (1995) to the absence of concentrated governance, prevailed. According to this logic, policy restrictions would be easier to adopt in systems where power is concentrated because partisans of the status quo in such systems are deprived of veto points. Where concentrated governance is absent, the exercise of a veto plays in favor of permissive policies. In contrast, the absence of concentrated governance played out in the restrictive direction when it combined with the presence of issue networks and administrative preferences matching those of biotechnology opponents. This was the case in Germany and Switzerland in both sectors and the United Kingdom and Sweden in the GMO sector. In these cases, the combination of conditions failed to encourage the exercise of a veto to maintain permissive policy designs. Rather, the absence of concentrated governance increased access to decision-making sites or facilitated venue shopping by actors opposing or concerned about biotechnology (Baumgartner and Jones 1993). These actors took advantage of the absence of concentrated governance to demand and obtain restrictive policy designs.

This finding differs considerably from existing analyses of biotechnology policy development, which suggests that variables exert policy influence in one specific direction. Bernauer and Meins (2003), for example, observe that the fragmentation of governance increases the number of access points to biotechnology opponents and therefore attribute a restrictive influence to this variable. Likewise, Young (2003) argues that policy transfers press for increases in the restrictiveness of GMO policies, independently of other factors. One of the key conclusions of this study is that concentrated governance, the mobilization of international rules and the presence or absence of conservative or Christian Democratic Party governments will play out differently in different combinations of conditions.

There is a second key difference between our findings and existing analyses of biotechnology policies. Considering only restrictive and permissive cases, we have six conditions explaining eleven cases. The six conditions span all three approaches, the policy network, country pattern and internationalization approaches, supporting the idea that policy explanations should draw from the three perspectives. Indeed, most of the existing analyses center on a single approach. For example, Montpetit (2005) and Toke and Marsh (2003) focus on policy networks. Vogel's (1986) focus on approaches to risk corresponds to a country pattern approach. Young (2003), who stresses the importance of policy transfers, falls under the internationalization approach. Sheingate (2006) may be the exception because he provides a study that integrates sector and country related factors, but he nevertheless leaves out international factors.

In the intermediary category, we have seven cases: Canada and Sweden in the ART sector; Belgium in the GMO sector; France and the Netherlands in both sectors. Explanations of these seven intermediate cases are complex since none of the conditions cut across the six equations (equations 3 to 8). In other words, more combinations of conditions lead to intermediate biotechnology policy than to either of permissive or of restrictive policies. Some patterns, however, emerge from these combinations. The cases of Canada in the ART sector and the cases of France, Belgium and the Netherlands in the GMO sector (equations 3, 4, and 7) suggest that intermediate biotechnology policy can be understood as the result of a compromise, or even a convergence of beliefs toward the middle through learning between targets and actors concerned about biotechnology (Sabatier and Jenkins-Smith 1999). In the cases of Canada (ART) and France (GMO), two conditions pull in different directions. On the one hand, an issue network of opposing or concerned actors pulls in the direction of restrictive policies. On the other hand, a policy community of targets pulls in the direction of permissive policies. In the ART sector in Canada and in the GMO sector in France, this combination encouraged the adoption of intermediate policies. In two cases, Belgium and the Netherlands in the GMO sector, the actors of the issue network also used European Union policies to press for restrictions. This advantage for opponents or concerned actors was however offset by the permissive preferences of an administrative agency in the Netherlands and the governing party in Belgium. This balance between permissive and restrictive conditions encouraged the adoption of the policy compromise represented by intermediate policy designs. Although not quite this well balanced, the equation explaining the French ART policy design (equation 8) is also one conducive to a compromise. The observation of compromises and possible learning within the biotechnology policy sectors is in itself an important observation, as it suggests that biotechnology controversies might not be as intractable as often suggested.

The results from the other intermediary cases, Sweden and the Netherlands in the ART sector, appear more puzzling. In fact, equations 5 and 6 do not present as clear a balance between permissive and restrictive leaning conditions as in the other intermediate cases. For example, the Swedish ART policy cannot be understood as a compromise between an issue network of biotechnology opponents and a policy community of targets, as both were absent. In fact, a permissive ART policy design would have made more sense for Sweden as the responsible administrative agency held preferences similar to those of target groups (see equation 5). Likewise, the presence of a community of targets and of an administrative agency whose preferences were favorable to targets should have led to a permissive ART design in the Netherlands (see equation 6). Why did these two countries end up with intermediate ART policies?

The answer to this question rests with concentrated governance, a condition present in the explanation of the Swedish and the Netherlands' ART policy design, as well as in the French, but notoriously absent from all the other equations in table 12.2. Institutional strands of policy studies have paid a particular attention to this question of concentrated governance, asking whether or not political institutions that locate power with the executive branch of government lead to the adoption of uncompromising policy alternatives (Weaver and Rockman 1993). In the biotechnology sector, one would intuitively think that concentrated governance affords policy-makers the risk of making either restrictive or permissive policy choices, avoiding the compromise embedded in the intermediate policy designs. From this perspective, the adoption of intermediate ART policy in the Netherlands and Sweden, in absence of conditions pressing for a compromise, is surprising. Paul Pierson's (1993) work helps to explain the moderation of decision-makers when they are in possession of full authority. With absolute power, he argues, also comes absolute responsibility. Systems that concentrate power also draw clear accountability lines, which encourage caution on the part of decision-makers.

Pierson's argument goes against much of the public choice literature, which is strongly suspicious of the discretionary use of authority in systems of power concentration (Ostrom 1990; Ostrom, Gardner, and Walker 1994). As Scharpf (1994: 32) puts it, public choice theory leads analysts to expect leaders in position of authority in hierarchical organizations to make "predatory" and even "oppressive" decisions. However, as discussed above, concentrated governance was absent in all cases where restrictive policies were adopted, the only policy output one could possibly associate with oppression. This result is consistent with observations made by Scharpf (1994: 37) in the studies of German hierarchical administrative organizations, which he conducted with Renate Mayntz. In systems where power concentration exists, leaders do not use their discretionary authority in decision-making to the extent predicted by public choice theory. While public choice theory suggests that vertical authority harms horizontal cooperation, Scharpf (1994; 1997) powerfully argues that the authority conferred upon actors at the summit of hierarchies merely casts a shadow conducive to cooperation among the actors at the bottom of these hierarchies. In other words, the mistake of public choice theory is to oppose vertical hierarchy and horizontal cooperation. Governance concentration is not inconsistent with the making of compromises and might even encourage it. The policy community of targets in the Netherlands' ART sector was inclined to accept some policy restrictions because concentrated governance did cast a shadow over it. Targets had an interest in accepting mild forms of restrictions because they knew policy-makers were capable of imposing much tougher ones on them. With similar knowledge about the capacity of policy-makers to impose decisions, the Swedish administrative agency, whose ART preferences coincided with those of targets, had an interest to accept some restrictions. Lastly, the concentration of governance in the French ART sector contributed to the policy compromise achieved in this country, even if the equation already presented a balanced set of conditions (equation 8). In short, concentrated governance appears only in equations associated with intermediate biotechnology policy designs, providing strong support to the idea that hierarchy and power concentration are consistent with policy moderation and compromises.

Table 12.2 does not associate policy designs in the ART and GMO sectors with clearly distinctive causal conjunctures. Some noticeable differences between the two sectors are nonetheless worth discussing. The biotechnology literature refers often to a so-called "Atlantic divide," which alludes to a European resistance and a North American acceptance of biotechnology. This divide is not so clear in the ART sector. In fact, two European countries, Belgium and the United Kingdom, have permissive policies while a North American country, Canada, has adopted an intermediate design in this same sector. The Atlantic divide is more apparent in the GMO sector. The presence of issue networks of actors opposing or concerned about biotechnology characterizes all European countries in this sector. In contrast, GMO issue networks were absent in the two North American countries (see equations 1, 2, 3, 4, 7 and 10). In all EU member states but France, the equations show that opponents and concerned actors used European Union regulations, which is consistent with much of the biotechnology policy literature. Vogel (2004), for instance, associates the restrictive nature of the European GMO policy to the Union's approach to risk. Bernauer and Meins (2003: 677) argue that the success of anti-GMO groups is not only attributable to the greater public outrage in Europe, but also to the multilevel policy environment provided by the European Union. In fact, any member state adopting a permissive GMO policy would be in violation of EU norms.

The impact of the European Union on member states' GMO policy, however, is, at best, only half of the story. Not only do differences exist in the GMO policy designs advocated by member states at the European level, there are considerable differences in the transposition of EU regulations in domestic law and in its implementation. Three European Union member states thus have restrictive GMO policies (Germany, United Kingdom and Sweden) and three have intermediate GMO policies (Belgium, the Netherlands and France). This suggests that domestic conditions cannot be neglected in policy analyses, even in sectors where the European Union might appear as the dominant actor. The equations relevant to ART further underline the importance of domestic conditions in explaining restrictive policy designs in the biotechnology sector, especially the presence of issue networks involving sympathetic administrative agencies. In other words, the success of biotechnology opponents in Europe should not be attributed too rapidly to a multilevel policy-making environment absent from North America.

In contrast to most policy studies, this Qualitative Comparative Analysis does not try to validate a single theoretically cohesive explanation. Instead, we have identified a number of conjunctures or combinations of conditions, drawn from three common policy-making approaches: the policy network, the country pattern and the internationalization approaches. Conditions associated with each of these approaches were present in one or the other of the eleven conjunctures that we identified, lending support to the idea that policy explanations need to draw from several theoretical approaches. Also distinct from most policy studies is the idea that the conjuncture matters more than the presence of conditions taken individually. In fact, in different combinations, the absence of concentrated governance notably played out differently, for example, encouraging permissiveness in the American ART sector but restrictiveness in Switzerland. At the substantive level, we found that issue networks and administrative preferences play a large role in biotechnology policy-making. In the GMO sector, the presence of issue networks made for restrictive policies in Europe while their absence had the reverse effect in North America. In addition, we found that concentrated governance does not have the effect predicted by public choice theory. In the ART sector of France, Sweden and the Netherlands, policymakers were more risk averse than their power required them to be, opting for the compromise solution that intermediate policies represent.


## CONCLUSION

This QCA analysis presents an understanding of biotechnology policy development that is more complex than those of previous analyses. Even if it confirms the divide between Europe and North America over GMO policy, it also shows that no such divide exists in the ART sector. In addition, the QCA reveals that the divide over GMO policy is better explained by the presence of issue networks and restrictive administrative preferences in several European countries than by European Union policies. This is not to say that Europeanization failed to exert policy influence in the biotechnology sector, but only that domestic factors prevailed most of the time. It should be recalled that several biotechnology policy studies attribute the restrictive European approach in the biotechnology sector to the European Union. In fact, the originality of this analysis rests with the idea that whether a condition enables or prevents a given policy output depends on the combination of conditions in which it is situated. For example, the reduction in the concentration of governance in several European countries, resulting from European integration, turned out to be an asset for biotechnology opponents, mobilized in domestic issue networks. In contrast, the absence of concentrated governance, historically glorified in North America, was an asset for biotechnology proponents. To fully capture the complexity of causality in policy studies, it is useful to draw from several theoretical perspectives and rely on a method that takes into consideration conjunctures or sets of conditions rather than individual variables into account.

Reality, however, is even more complex than we suggest in this analysis. In this study, we focused our attention on policy-making at the domestic level and found that actors who were concerned about or opposed to biotechnology more frequently resorted to international rules to press for the adoption of their preferred policy options than target groups did. This finding should be surprising to anyone who believed that internationalization first and foremost favored multinational firms, whose presence is important in the GMO sector. It should be no less surprising to political economists who argue that the international trade regime, as shaped by the World Trade Organization, severely constrains the adoption of policies that are impediments to the free movement of goods and services across borders. If international rules did not play the role several analysts expected of them, it may be that we have insufficient knowledge about their construction. Therefore, future biotechnology analyses should place greater attention on the impact of domestic factors on the production of international rules. International relations scholars such as Young (2003), Coleman and Gabler (2002), Vogel (1986), and Tiberghien and Starrs (2004) have already begun research on the construction of international rules for biotechnology. Policy scholars should join their efforts to provide a better understanding of the interplay between domestic sectoral biotechnology networks on the one hand and the international politics of biotechnology on the other hand.

## NOTE

1. We are grateful to Sakura Yamasaki for methodological assistance in doing the Qualitative Comparative Analysis.

## REFERENCES

Bachrach, P., and M. S. Baratz. 1963. Decisions and Nondecisions: An Analytical Framework. *American Political Science Review* 57 (3): 632–42.

Baumgartner, F., and B. Jones. 1993. *Agendas and Instability in American Politics*. Chicago: University of Chicago Press.

Bernauer, T., and E. Meins. 2003. Technological Revolution Meets Policy and the Market: Explaining Cross-National Differences in Agricultural Biotechnology Regulation. *European Journal of Political Research* 42 (5): 643–83.

Bleiklie, I., M. Goggin, and C. Rothmayr (eds.). 2004. *Governing Assisted Reproductive Technology: A Cross-Country Comparison*. London: Routledge.

Coleman, W. D., and M. Gabler. 2002. Agricultural Biotechnology and Regime Formation: A Constructivist Assessment of the Prospects. *International Studies Quarterly* 46 (4): 481–506.

Cronqvist, L. 2006. TOSMANA. Tool for Small-N Analysis (SE Version 1.25). www.tosmana.net

De Meur, G., and D. Berg-Schlosser. 1996. Conditions of Authoritarianism, Fascism, and Democracy in Interwar Europe: Systematic Matching and Contrasting of Cases for "Small N" Analysis. *Comparative Political Studies* 29 (4): 423–68.

De Meur, G., and B. Rihoux. 2002. *L'analyse quali-quantitative comparée*. Louvain-la-Neuve: Academia Bruylant.

Montpetit, É. 2005. A Policy Network Explanation of Biotechnology Policy Differences between the United States and Canada. *Journal of Public Policy* 25: 339–66.

Ostrom, E. 1990. *Governing the Commons: The Evolution of Institutions for Collective Action*. Cambridge: Cambridge University Press.

Ostrom, E., R. Gardner, and J. Walker. 1994. *Rules, Games & Common-Poll Resources*. Ann Arbor: The University of Michigan Press.

Pierson, P. 1993. *Dismantling the Welfare State? Reagan, Thatcher, and the Politics of Retrenchment*. Cambridge: Cambridge University Press.

Przeworski, A., and H. Teune. 1970. *The logic of comparative social inquiry*. New York: Wiley.

Ragin, C. C. 1987. *The Comparative Method: Moving Beyond Qualitative and Quantitative Strategies*. Berkeley: University of California Press.

Sabatier, P. A., and H. C. Jenkins-Smith. 1999. The Advocacy Coalition Framework: An Assessment. In *Theories of the Policy Process*. Ed. Paul A. Sabatier. Pp. 117–66. Boulder: Westview Press.

Scharpf, F. W. 1994. Positive und negative Koordination in Verhandlungssystemen. *PVS-Sonderheft 24, Policy-Analyse. Kritik und Neuorientierung*: 57–83.

———. 1997. *Games Real Actors Play: Actor-Centered Institutionalism in Policy Research*. Boulder: Westview Press.

Sheingate, A. D. 2006. Promotion versus Precaution: The Evolution of Biotechnology in the United States. *British Journal of Political Science* 36: 243–68.

Tiberghien, Y., and S. Starrs. 2004. The EU as Global Trouble-Maker in Chief: A Political Analysis of EU Regulations and EU Global Leadership in the Field of Genetically Modified Organisms. Prepared for the 2004 Conference of Europeanists, the Council of European Studies.

Toke, D., and D. Marsh. 2003. Policy Networks and the GM Crops Issue: Assessing the Utility of a Dialectical Model of Policy Networks. *Public Administration*. 81 (2): 229–51.

Tsebelis, G. 1995. Decision making in political systems: veto players in presidentialism, parliamentarism, multicameralism and multipartyism. *British Journal of Political Science* 25 (3): 289–325.

Varone, F., C. Rothmayr, and É. Montpetit. 2006. Regulating Biomedicine in Europe and North America: A Qualitative Comparative Analysis. *European Journal of Political Research* 45 (2): 317–43.

Vogel, D. 1986. *National Styles of Regulation: Environmental Policy in Great Britain and the United States.* Ithaca and London: Cornell University Press.

———. The Hare and the Tortoise Revisited: The New Politics of Risk Regulation in Europe and the United States. In *Transatlantic Policymaking in an Age of Austertity: Diversity and Drift.* Eds. M. Levin. and M. Shapiro. Pp. 177–202. Washington: Georgetown University Press.

Young, A. R. 2003. Political Transfer and "Trading Up"? Transatlantic Trade in Genetically Modified Food and U.S. Politics. *World Politics* 55: 457–84.

Weaver, K., and B. A. Rockman (eds.). 1993. *Do Institutions Matter? Government Capabilities in the United States and Abroad.* Washington: The Brookings Institution.

# Appendix 12.1: Hypotheses and their Translation for QCA

| *Hypotheses as formulated in chapter one* | *Hypotheses translated in QCA format* |
|---|---|
| H1: When target groups belong to a policy community, understood as a closed network of civil society and state actors who share similar beliefs, biotechnology policies should be permissive. | - Policy community of targets (pct) → permissive policy (code 0)*<br>- Absence of a policy community of targets (PCT) → restrictive policy (code 1) |
| H2: Where issue networks exist, groups opposed to or concerned by potential negative effects of biotechnology are more successful at exerting pressures favorable to restrictive biotechnology policies. | - Issue network of opponents (INO) → restrictive policy (code 1)<br>- Absence of an issue network of opponents (ino) → permissive policy (code 0) |
| H3: When a Christian and/or neo-conservative (or green in the GMO case) controls cabinet positions with biotechnology responsibilities, policies should be more restrictive. Conversely, if secular and/or progressive parties govern, biotechnology policies should be more permissive. | - Conservative party label (CPL) → restrictive policy (code 1)<br>- Absence of conservative party label (cpl) → permissive policy (code 0)<br>- Progressive party label (ppl) → permissive policy (code 0)<br>- Absence of progressive party label (PPL) → restrictive policy (code 1) |
| H4: When groups opposed to biotechnologies are involved in various policy-making arenas, and if those arenas can coordinate their policy-making activities, biopolicy decisions should be restrictive towards target groups. Conversely, when target groups can confine policy-making to a single site of power, or if coordination between the various policy-making arenas is weak enough to encourage regulatory competition, then the content of the policy should be more permissive. | - Fragmented governance with coordination (FGC) → restrictive policy (code 1)<br>- Absence of fragmented governance (fgc) → permissive policy (code 0)<br>- Fragmented governance without coordination (fgn) → permissive policy (code 0)<br>- Absence of fragmented governance (FGN) → restrictive policy (code 1)<br>- Concentrated governance (cog) → permissive policy (code 0)<br>- Absence of concentrated governance (COG) → restrictive policy (code 1) |

H5: When the preferences of administrative agencies responsible for biotechnology are jibing with those of target groups, policies should be permissive. Conversely, when the preferences of administrative agencies responsible for biotechnology are jibing with those of opponents or groups concerned about biotechnology, policies should be restrictive.

- Administrative preferences jibing with targets (apt) → permissive policy (code 0)
- Absence of administrative preferences jibing with targets (APT) → restrictive policy (code 1)
- Administrative preferences jibing with opponents (APO) → restrictive policy (code 1)
- Absence of administrative preferences jibing with opponents (apo) → permissive policy (code 0)

H6: When opponents or groups concerned with biotechnology are successful at using international or European norms, policies should be restrictive. Conversely, when target groups are successful at using international or European norms, policies should be permissive.

- International rules used by opponents (IRO) → restrictive policy (code 1)
- Absence of international rules used by opponents (iro) → permissive policy (code 0)

H7: When target groups are able to demonstrate the effectiveness of permissive solutions adopted in other countries, as well as the pernicious effects of more restrictive policies, biotechnology policy should be permissive Conversely, when opponents or groups concerned with biotechnology can draw from foreign experiences with restrictive policies that work, biotechnology policy should be more restrictive.

- International rules used by targets (irt) → permissive policy (code 0)
- Absence of international rules used by targets (IRT) → restrictive policy (code 1)
- Policy transfer initiated by opponents (PTO) → restrictive policy (code 1)
- Absence of policy transfer initiated by opponents (pto) → permissive policy (code 0)
- Policy transfer initiated by targets (ptt) → permissive policy (code 0)
- Absence of policy transfer initiated by targets (PTT) → restrictive policy (code 1)

*Abbreviations in uppercase letters suggest an influence toward restrictive policies, coded 1. Abbreviations in lowercase suggest an influence toward permissive policies, coded 0.

## Appendix 12.2 Truth Tables

### Restrictive cases

| CASES* | PCT | INO | CPL | PPL | FGC | FGL | COG | APT | APO | IRO | IRT | PTO | PTT | OUT |
|---|---|---|---|---|---|---|---|---|---|---|---|---|---|---|
| GA,CHA,CHG | 1 | 1 | 1 | 0 | 1 | 1 | 1 | 1 | 1 | 0 | 1 | 0 | 1 | 1 |
| GG | 1 | 1 | 0 | 0 | 1 | 1 | 1 | 1 | 1 | 1 | 1 | 0 | 1 | 1 |
| CA | 0 | 1 | 0 | 1 | 0 | 0 | 1 | 1 | 1 | 0 | 1 | 0 | 1 | 0 |
| CG,USAG | 0 | 0 | 0 | 1 | 0 | 1 | 1 | 0 | 0 | 0 | 1 | 0 | 1 | 0 |
| USAA | 1 | 0 | 0 | 1 | 0 | 0 | 1 | 1 | 0 | 0 | 1 | 0 | 1 | 0 |
| UKA | 0 | 0 | 0 | 1 | 0 | 1 | 0 | 1 | 0 | 0 | 1 | 0 | 1 | 0 |
| UKG,SWG | 1 | 1 | 0 | 1 | 0 | 1 | 1 | 1 | 1 | 1 | 1 | 0 | 1 | 1 |
| FA | 1 | 1 | 1 | 1 | 0 | 1 | 0 | 1 | 0 | 0 | 1 | 0 | 1 | 0 |
| FG | 0 | 1 | 0 | 1 | 1 | 1 | 1 | 1 | 0 | 1 | 1 | 0 | 1 | 0 |
| SWA | 1 | 0 | 0 | 0 | 0 | 1 | 0 | 1 | 1 | 0 | 1 | 0 | 1 | 0 |
| NLA | 0 | 0 | 1 | 0 | 0 | 1 | 0 | 0 | 0 | 0 | 1 | 0 | 1 | 0 |
| NLG | 1 | 1 | 0 | 1 | 1 | 1 | 1 | 0 | 0 | 1 | 1 | 0 | 1 | 0 |
| BA | 0 | 0 | 1 | 1 | 0 | 0 | 1 | 0 | 0 | 0 | 1 | 0 | 1 | 0 |
| BG | 1 | 1 | 1 | 0 | 1 | 1 | 1 | 1 | 1 | 1 | 1 | 0 | 1 | 0 |

### Intermediate Cases

| CASES* | PCT | INO | CPL | PPL | FGC | FGL | COG | APT | APO | IRO | IRT | PTO | PTT | OUT |
|---|---|---|---|---|---|---|---|---|---|---|---|---|---|---|
| GA,CHA,CHG | 1 | 1 | 1 | 0 | 1 | 1 | 1 | 1 | 1 | 0 | 1 | 0 | 1 | 0 |
| GG | 1 | 1 | 0 | 0 | 1 | 1 | 1 | 1 | 1 | 1 | 1 | 0 | 1 | 0 |
| CA | 0 | 1 | 0 | 1 | 0 | 0 | 1 | 1 | 1 | 0 | 1 | 0 | 1 | 1 |
| CG,USAG | 0 | 0 | 0 | 1 | 0 | 1 | 1 | 0 | 0 | 0 | 1 | 0 | 1 | 0 |
| USAA | 1 | 0 | 0 | 1 | 0 | 0 | 1 | 1 | 0 | 0 | 1 | 0 | 1 | 0 |
| UKA | 0 | 0 | 0 | 1 | 0 | 1 | 0 | 1 | 0 | 0 | 1 | 0 | 1 | 0 |
| UKG,SWG | 1 | 1 | 0 | 1 | 0 | 1 | 1 | 1 | 1 | 1 | 1 | 0 | 1 | 0 |
| FA | 1 | 1 | 1 | 1 | 0 | 1 | 0 | 1 | 0 | 0 | 1 | 0 | 1 | 1 |
| FG | 0 | 1 | 0 | 1 | 1 | 1 | 1 | 1 | 0 | 1 | 1 | 0 | 1 | 1 |
| SWA | 1 | 0 | 0 | 0 | 0 | 1 | 0 | 1 | 1 | 0 | 1 | 0 | 1 | 1 |
| NLA | 0 | 0 | 1 | 0 | 0 | 1 | 0 | 0 | 0 | 0 | 1 | 0 | 1 | 1 |
| NLG | 1 | 1 | 0 | 1 | 1 | 1 | 1 | 0 | 0 | 1 | 1 | 0 | 1 | 1 |
| BA | 0 | 0 | 1 | 1 | 0 | 0 | 1 | 0 | 0 | 0 | 1 | 0 | 1 | 0 |
| BG | 1 | 1 | 1 | 0 | 1 | 1 | 1 | 1 | 1 | 1 | 1 | 0 | 1 | 1 |

### Permissive Cases

| CASES* | PCT | INO | CPL | PPL | FGC | FGL | COG | APT | APO | IRO | IRT | PTO | PTT | OUT |
|---|---|---|---|---|---|---|---|---|---|---|---|---|---|---|
| GA,CHA,CHG | 1 | 1 | 1 | 0 | 1 | 1 | 1 | 1 | 1 | 0 | 1 | 0 | 1 | 0 |
| GG | 1 | 1 | 0 | 0 | 1 | 1 | 1 | 1 | 1 | 1 | 1 | 0 | 1 | 0 |
| CA | 0 | 1 | 0 | 1 | 0 | 0 | 1 | 1 | 1 | 0 | 1 | 0 | 1 | 0 |
| CG,USAG | 0 | 0 | 0 | 1 | 0 | 1 | 1 | 0 | 0 | 0 | 1 | 0 | 1 | 1 |
| USAA | 1 | 0 | 0 | 1 | 0 | 0 | 1 | 1 | 0 | 0 | 1 | 0 | 1 | 1 |
| UKA | 0 | 0 | 0 | 1 | 0 | 1 | 0 | 1 | 0 | 0 | 1 | 0 | 1 | 1 |
| UKG,SWG | 1 | 1 | 0 | 1 | 0 | 1 | 1 | 1 | 1 | 1 | 1 | 0 | 1 | 0 |
| FA | 1 | 1 | 1 | 1 | 0 | 1 | 0 | 1 | 0 | 0 | 1 | 0 | 1 | 0 |
| FG | 0 | 1 | 0 | 1 | 1 | 1 | 1 | 1 | 0 | 1 | 1 | 0 | 1 | 0 |
| SWA | 1 | 0 | 0 | 0 | 0 | 1 | 0 | 1 | 1 | 0 | 1 | 0 | 1 | 0 |
| NLA | 0 | 0 | 1 | 0 | 0 | 1 | 0 | 0 | 0 | 0 | 1 | 0 | 1 | 0 |
| NLG | 1 | 1 | 0 | 1 | 1 | 1 | 1 | 0 | 0 | 1 | 1 | 0 | 1 | 0 |
| BA | 0 | 0 | 1 | 1 | 0 | 0 | 1 | 0 | 0 | 0 | 1 | 0 | 1 | 1 |
| BG | 1 | 1 | 1 | 0 | 1 | 1 | 1 | 1 | 1 | 1 | 1 | 0 | 1 | 0 |

*The cases are identified by the abbreviation of the country, as indicated in Table 12.1, followed by the letter A for ART and G for GMO.

# INDEX

# ABOUT THE CONTRIBUTORS

**Gabriele Abels** is a Senior Research Fellow at the Institute for Science and Technology Studies and a Professor of Comparative Politics at Bielefeld University. In 2005, she was a visiting Professor at University of Missouri-St. Louis. She has a Ph.D. from the University of Essen (1999). She works on European integration, biotechnology policy in the European Union and in Germany, and on technology assessment and gender issues. She is the coeditor of "femina politica," a German feminist political science journal, as well as the coauthor of *Demokratische Technikbewertung* (transcript Verlag). She has published in *European Integration Online Papers*, *Science as Culture*, *Österreichische Zeitschrift für Poliktikwissenschaft* and the *Encyclopedia of Sociology*.

**Francis Garon** holds a Ph.D. in political science from the Université de Montréal (2006). He specializes in public administration and public policy. More specifically, his work is on public participation in the policy process in Canada. He is currently a postdoctoral researcher, assessing the contribution of community associations to Quebec's policy in twenty sectors. He published an article in *Éthique Publique* in 2005 and coauthored a number of book chapters.

**Éric Montpetit** is a Professor of Political Science at the Université de Montréal. He holds a Ph.D. from McMaster University (1999) and was a fellow of the Fulbright program at Duke University in North Carolina (1997–1998). His work is on environmental and biotechnological policy development in North America and Europe. He is the author of *Misplaced Distrust* (UBC Press), winner of the 2006 American Political Science Association's Lynton Keith Caldwell Prize for the best book on environmental politics and policy. He also published notably in *World Politics*, *Governance*, the *Journal of Public Policy*, the *Journal of European Public Policy*, *Comparative Political Studies*, and *Policy Sciences*.

**Christine Rothmayr** is Professor of Political Science at the University of Montreal, Canada. She has previously taught at the University of Geneva in Switzerland. She holds a Ph.D. in political science from University of Zurich (1999). Her field of interest is comparative public policy, where she studies policies in the field of biotechnology, biomedicine and higher education. Her other two fields of activity are courts and politics in comparative perspective and public opinion and public policy. Her work has been published in the *European Journal of Political Research*, *Comparative Political Studies*, *International Journal of Public Opinion Research*, and *West European Politics*.

**Francesca Scala** is an Associate Professor at the Department of Political Science at Concordia University in Montreal, Canada. She received her Ph.D. in public policy from the School of Public Policy and Administration at Carleton University (2002). Her areas of specialization include biotechnology policy, and citizen engagement from both a Canadian and comparative perspective. Dr. Scala's work has been published in a number of edited volumes and journals, including *Policy Sciences*, *Society and Natural Resources*, the *Canadian Journal of Political Science*, and *Policy and Society*.

**Nathalie Schiffino** is a Political Science Lecturer at the Université de Louvain (UCL) and the Université de Mons (FUCaM). She holds a Ph.D. from the Université de Louvain (2000). Her work focuses on biotechnological policy development, specifically biomedicine and genetic engineering in Europe. Her other main area of research is the management of political crises in democratic regimes. She is the author of *Crises politiques et démocratie en Belgique* (L'Harmattan, Coll. Logiques politiques). She also authored and coauthored several articles, notably in *Archives of Public Health*.

**Arco Timmermans** is a Professor at the Institute for Governance Studies at the University of Twente in the Netherlands. He received his Ph.D. in 1996 from the European University Institute in Florence. His research interests include comparative public policy, with a focus on the role of institutions in decision-making. He has published in several edited books and scholarly journals, including the *European Journal of Political Research* and the *Journal of Comparative Policy Analysis*.

**Frédéric Varone** is a political science professor at the Université de Genève. He taught previously at the Université catholique de Louvain (Belgium, 1999-2005). He holds a Ph.D. from the University of Bern (1998). His work is on comparative policy analysis, program evaluation and reforms of the public sector. He published notably in the *Canadian Journal of Political Science, Comparative Political Studies, European Journal of Political Research, Evaluation, International Review of Administrative Sciences, Journal of European Public Policy, Public Management Review* and *Revue Internationale de Politique Comparée*.